FORD'S GOLDEN FIFTIES

FORD'S GOLDEN FIFTIES

Ford Rotunda, Dearborn, Michigan, 1936-62

BY LORIN SORENSEN

Left: *A '51 Ford Custom Convertible Coupe at the Rouge plant in Dearborn, Michigan, place of its birth; from raw ore and hot steel, to heart-stirring beauty.*

SILVERADO PUBLISHING COMPANY, ST. HELENA, CALIFORNIA
and Ten Speed Press, Berkeley / Toronto

Copyright © 2003, 1997 by Lorin Sorensen

All rights reserved. No part of this book may be reproduced in any form, except brief excerpts for the purpose of review, without written permission of the publisher.

Published by Silverado Publishing Company
P.O. Box 393
St. Helena, California 94574

Available from Ten Speed Press
P.O. Box 7123
Berkeley, California, 94707
www.tenspeed.com

Distributed in Australia by Simon & Schuster Australia, in Canada by Ten Speed Press Canada, in New Zealand by Southern Publishers Group, in South Africa by Real Books, and in the United Kingdom and Europe by Airlift Book Company.

Edited by Peggy Noonan
Typesetting by Ellen Peters
Cover art by Joseph Maas

ISBN NO. 1-58008-550-4
Printed in China

First printing, 1997
1 2 3 4 5 6 7 8 9 10 – 08 07 06 05 04 03

Cover Photo: A 1951 Ford Custom *Convertible Coupe* comes off the line at Dearborn Assembly.

ACKNOWLEDGEMENTS

A very special thanks to my good friend Souren Keoleian who spent 35 years with Ford and is still one of the most enthusiastic Ford "buffs" I've ever met. And to Lee Kollins, auto historian, who was with Ford for 44 years and never fails to tell a good story about his adventures in Public Relations and the high offices of Ford World Headquarters.

Also, thanks to Dan Brooks who loaned some great photos for this book and to all of the Ford fans and friends who have given me encouragement through the years.

Much appreciation to my friends Dave Crippen and Mike Davis who, for the Henry Ford Museum, conducted the "Oral Reminiscences" interviews with the Ford stylists and others quoted in this book.

Also, thanks to Dan Erickson and Rosemarie Stanish of Ford Photographic, and to Linda Skularis and others at the Henry Ford Museum Research Center.
LORIN SORENSEN ◆

— CONTENTS —

	INTRODUCTION	9
1949	CAR OF THE YEAR	11
1950	A CAR MAKER'S DREAM	25
1951	AERO-STYLED BEAUTY	43
1952	THE NEW CUSTOM LOOK	58
1953	FORD AT FIFTY	69
1954	A MILESTONE V-8	85
1955	FORD GOES SPORTY	104
1956	YEAR OF THE CLASSICS	129
1957	AMERICA'S FAVORITE	157
1958	THE THUNDERBIRD LOOK	181
1959	FORD AT FIFTY-MILLION	195

A classic fifties shot taken at the Ford Styling Rotunda reflecting pool in Dearborn to show how America's sweethearts and the sporty new '56 Thunderbird were meant for each other.

Left: One of the best-selling Ford models of all time, a handsome 1956 Fairlane Victoria has its portrait taken in front of its namesake "Fairlane", Henry and Clara Ford's old estate home in Dearborn, Michigan.

Ford stylist George Walker, left, Edsel's son William Clay Ford, and Dick Krafve, manager of the Edsel project, at the Edsel Styling Studios in 1957. Walker was the man most responsible for Ford car designs throughout the "golden fifties."

INTRODUCTION

"THE WALKER CARS"

THIS is a book about some of those great cars from "Ford's Golden Fifties" which were produced as the result of decisions made almost exclusively by two men — company president Henry Ford II, who had the last word on all the designs, and George W. Walker, his consultant.

The role Walker played in Ford styling during that decade is not fully understood, yet his influence on the '49-59 models is so clear they might be called "the Walker cars."

That's Walker in the photo opposite, with his devilish trademark grin, leaning in on an annoyed William Clay Ford a while after Henry II had named him the first "Vice-President and Director of Styling" in the company's history. The man standing is just as uncomfortable as young Ford, and more so. He is R.E. "Dick" Krafve who headed the Special Products Division, charged with the unenviable task of creating the "E" car, or Edsel that we see being designed in the background.

The two men were cool to the wheeler-dealer Walker being top gun in the styling department, but in the eyes of Henry Ford II who liked people who could deliver, nobody at Ford styling came close to the genius of George Walker!

Walker was one-of-a-kind . . . something like a P.T. Barnum who could draw. Born in Chicago in 1896, the son of a railroad conductor and a quarter-Cherokee Indian mother, he played professional football until he was 27 as a half-back under the great Jim Thorpe for the Cleveland Panthers. Then, he went to art school and became a fashion illustrator before moving to Detroit in 1929 where he opened the George W. Walker industrial design company. He soon found he had a talent that paid even better than drawing — the gift of sales and showmanship.

By the beginning of World War II he had many national accounts, designing everything from toasters to tractors. In the auto industry he was known best for his famous Packard "Clipper" and in 1946 picked up the Ford account when Henry Ford II was desperate for a '49 model that would beat Chevrolet in the sales race.

Walker turned in the winning '49 Ford design over the company's own designers and on the unprecedented success of that car began a long association with Henry Ford II as his outside styling consultant.

Walker's artistic eye quickly set the modern tone that helped Ford back toward the top. He went on to style all the '50s Lincolns, the Mercury from 1950, and such designs as the Thunderbird and the '52 Ford, which gave the company the round rocket taillights that would become a Ford trademark. Others would take credit for coming up with brilliant styling ideas but in the end it was Walker, the artistic, flamboyant, persuasive outside salesman who put them all together in a package and sold his friend Henry Ford II on the final designs.

In the summer of 1955 Henry Ford II was so indebted to Walker, especially after the brilliance of his Thunderbird, that he finally persuaded him to come to work for the company full time as VP of styling. Walker gave the company president something he had always wanted — a man to head his styling department with style — a man like GM's legendary Harley Earl.

George wouldn't let him down. At Ford he became a great glad-handing personality, a flashy dresser who owned 40 pairs of shoes and 70 suits, who always smelled of cologne, had a perpetual tan, a bent for silver dollar-size cuff-links, was handy with the ladies — and always got the job done turning out beautiful, top-selling Fords.

Befitting his new position as head of a staff of 650 artists, draftsmen, modelers and engineers he had his big office suite at the Ford styling center done in creamy-white and black, with raw silk draperies, sumptuous leather couches, with a jungle of tropical plants along one end. On the floor was spread an inch-thick carpet of black lambskin.

Said Walker happily when showing it to visitors: "Ain't it sexy!"

In 1957 Walker was such a hot Ford stylist that he made the cover of *Time* magazine, which called him *"The Cellini of Chrome"*.

But no one at Ford could say they ever saw him draw.

Gene Bordinat, who became his successor in 1961, understood Walker's talent for coaching others to get the result he wanted better than most: "Nobody paid him to draw lines. They paid him to come up with a car, and he came up with a car. He insisted that was the car the company should have, and it's the car that went out and did great work."

That was George Walker — a man who did great work. To him as a Ford stylist, above all the others — and, of course to Henry Ford II who liked what Walker rolled out — we say thanks for all the terrific cars! *LORIN SORENSEN*

1949 CAR OF THE YEAR

THE family-owned Ford Motor Company was fortunate there were sons to carry on the business following the death of its president Edsel Ford in 1943 and his father Henry Ford, the founder of the giant auto firm, in 1947.

Edsel's 26-year-old son Henry Ford II had come home from the Navy to take charge of the company during World War II and would soon prove himself by recruiting a top management team to get the company back on track as it resumed civilian production. Later, his younger brothers, Benson and William Clay Ford, would also play roles in the company.

Young Henry's biggest post-war challenge was how to revitalize the company and make it profitable again. Considering the few options, his decision was to put everything into a single best-selling new Ford model for 1949. The company's chief stylist Eugene "Bob" Gregorie was ordered to come up with its design in time for the car's introduction in mid-1948. Gregorie was a top stylist with years of Ford experience, having designed some of the pre-war V-8s, the first Mercury, and Edsel Ford's classy Continentals.

But, from the beginning, Gregorie was uncomfortable about designing the car that was so vital to the company's future. Henry II didn't have the knack to give him guidance, unlike the old days when his father — the articulate Edsel with the natural eye — would sit long hours in his studio looking over his shoulder making suggestions until they had exactly what they wanted.

So, Gregorie's '49 Ford design went along without clear direction.

It was soon mid-1946, with just three months to go before the car's design had to be set and drawings started for tooling. The Ford product planners were working up the production details when they suddenly came to the realization that Gregorie's car was just too big and costly to produce at a profit in the high-volume, low-price, sales competition with Chevrolet and Plymouth.

The only choice was to go back to the drawing board.

Henry II was stunned! He decided that the only quick solution was to bring in an outside styling consultant with some fresh ideas. Luckily, he chose well-known Detroit industrial designer George W. Walker whose automobile credentials included work on the Nash and the design of the Packard Clipper.

The resourceful Walker took on the project at his George Walker Design studios and within the astonishingly short time of just three months made a presentation to Henry of a radically-designed smooth-sided car model based on some original ideas from his staff, notably Dick Caleal, Bob Bourke, Holden Koto, and Joe Oros.

Gregorie's own Ford team also competed on the '49 Ford design but lost to Walker. However, while he resented this decision and resigned, Gregorie's work was not lost. His Ford model became the beautiful new '49 Mercury and his Mercury became the '49 Lincoln.

Walker accurately predicted that his revolutionary new '49 Ford design would begin a new styling trend in the industry... " There was a lot more significance in the 1949 Ford than the fact that it was different. It had to be. But more than that, it provided the basic concept of our styling since that time. Practically all cars at that time had bulging side lines, particularly around the front and rear fenders. We smoothed those lines out and began the movement toward the integration of the fenders and body."

Joe Oros, who worked for Walker, had a big hand in coming up with the '49 Ford's distinctive front-end theme. He remembers that the idea came from the effect airplane styling had on automobile men. "Without realizing it, we had set... some of the first aerodynamic considerations for an automobile which was the aircraft spinner theme..."

Within weeks of Henry Ford II's approval of the '49 Ford design, Walker's sketches and scale models jelled in the form of a handmade car. Then engineers slapped an old Ford body on the new chassis for camouflage and test-

Left: **Company president Henry Ford II, with his brothers Benson, left, and William Clay Ford, center, pose with a model of the new '49 Ford designed by outside consultant George W. Walker and his staff. In the marketplace this famous car saved the company financially and made Walker Ford's top stylist of the fifties.**

drove it through the streets of Detroit. With little time to spare the design was frozen and blueprints were sent to the Ford tool room. Best seller or lemon, the new Ford was set.

In record time, the first one rolled off the Dearborn production line April 12, 1948, a month after the new Lincolns and Mercurys, giving the company a nice jump on the competition.

The nation's press got their first "sneak preview" of the new Fords on June 8, 1948, at New York's Waldorf Astoria Hotel the day before the public was invited in for an extravagant week-long introductory show. Six bars were set up and champagne flowed like water as the reporters mobbed the new Fords. "It was like the good old days of prewar auto shows," wrote a veteran journalist from *Business Week*, "colorful decorations, company brass in a receiving line, name orchestra playing dance music and sales blurbs over loudspeakers."

The next day at the official public opening, thousands of

Models hired to present the new '49 Fords at New York's Waldorf Astoria Hotel in mid-June, 1948, get their first backstage look at the sporty new *Convertible*.

Left: A beautiful new '49 Ford *Convertible* in luscious *Miami Cream* goes by on the merry-go-round at the Waldorf introduction. Sparing no expense, Ford hired famed American designer Walter Dorwin Teague to create the fabulous show.

ordinary citizens crowded into the Waldorf's immense gold-and-white grand ballroom to see the cars. Amidst the sounds and glitter of the spectacle they saw revolving overhead on a modernistic 15-ton rotisserie, two-full size Ford cutaway chassis finished in copper, brass and bronze. On the floor, a giant merry-go-round 50-feet across carried five of the sensational new '49 Fords with beautiful, smiling girls waving to the crowd. And everywhere, in special exhibits around the ballroom and off to the side, narrators were explaining the features and letting people sit in the all-new Fords, Mercurys and Lincolns.

Moving around through the crowd a '49er "miner" handed out "Ford" imprinted plastic gold nuggets and special guests were given souvenir toy models of the new cars.

Throughout the event, as search lights lit the skies, Ford's spectacular Waldorf car show was covered by radio, television, and newsreels. A blimp with a lighted *"FORD"* in neon on its sides cruised overhead. Downtown on Allen's Alley, radio personality Fred Allen strolled with a microphone asking people what they thought of the new '49 Ford. By now, just about everyone in New York had seen it, either in person or in advance ads on thousands of billboards.

The following week the cars were unveiled at one of the biggest dealer introductions in American history – certainly the biggest for Ford since the 1928 Model A. It created an avalanche of orders, making the kickoff of the new Ford even more successful than Henry Ford II had ever hoped.

It was a multi-million dollar gamble that paid off! By the end of 1949, Ford Motor Company would pass the million mark for passenger car sales, including Lincoln and Mercury, and post huge profits. Yet, Henry Ford II would still not realize his own personal goal – which was to beat Chevy. The wily competitor also had a terrific year, leaving Ford in third place in the sales race, behind Plymouth. ◆

"... The thing that really counts is a well-designed product that millions of hands will stretch out to buy because it says something to them that they cannot resist. When the right lines and contours are applied to a product, they must evoke that particular emotion in that buyer..." GEORGE WALKER, FORD STYLIST

"I must say that I was rather shocked, (at Walker's design) primarily, because it was the first time that the so-called slabsided design had been implemented in three dimension ... Naturally, from a designer's point of view, it was quite exciting. GENE BORDINAT, FORD STYLIST

A *Tudor* body meets its V-8 chassis. Radical new '49 Ford features included a suspension system with independent front coil springs. Buyers had the choice of the 100 hp V-8 engine pictured, or the 95 hp "Six", and a new optional Overdrive.

Right: Film star Basil Rathbone and friends visit the Ford Dearborn assembly plant in the summer of 1948. Best known for his role as Sherlock Holmes, the versatile English actor is pictured behind the wheel of one of the plant's freshly-built '49 Ford convertibles.

Left: A '49 Ford *Convertible* comes off the line at Dallas in *Fez Red,* a color only available that year on the soft top models.

THE NEW 1949 FORDS

The 1949 Ford, which reveals a radical departure from traditional Ford styling and engineering, was made public today by the Ford Motor Company.

"New standards of beauty and comfort, economy and performance in the 1949 Ford passenger cars advance them far ahead of others in the low-priced field," J.R. Davis, vice-president and director of sales and advertising, said. "Styling of the new Ford definitely establishes it as the car of the year."

To develop and produce the 1949 Ford passenger cars, Ford Motor Company expended more than $37,400,000 in tools, dies, jigs and fixtures.

The 1949 design has been molded along functional lines, resulting in a long, low, sweeping silhouette. The grille is distinctive, the hood massive but shorter and the body so wide the rear fenders have been eliminated. There are clean, unbroken lines from front to rear.

There are two lines of cars, the Ford and the Ford Custom. Body styles in both lines include the *four-door Sedan, two-door Sedan,* and *Club Coupe. Convertible* and *Station Wagon* models are obtained only in the Custom line and the *three-passenger Coupe* only in the Ford line.

There are eight durable new exterior colors – *Bayview Blue Metallic, Birch Gray, Sea Mist Green, Arabian Green, Colony Blue, Gun Metal Gray Metallic, Midland Maroon Metallic,* and *Black*.

Two additional colors – *Fez Red* and *Miami Cream* – are available in the *Convertible* only. FORD NEWS BUREAU, DEARBORN, JUNE 9, 1948

Like a scene from a Norman Rockwell painting, a typical crowd of working-class Americans look over the new '49 Ford at a Portland, Oregon dealer showroom. A lumberjack working in the woods made about $2.50 an hour and could have a new *Tudor* V-8 like this one for about $1,505 F.O.B. Detroit.

Right: A prospective customer at Bishop Ford in Santa Rosa, California lets his wife try the wheel of a new '49 Custom Deluxe *Convertible Coupe,* as the wily salesman looks on. Women were taking a larger role in buying the family car.

1949 was a seller's market. Said one astonished Ford dealer:

"A guy comes in and he looks at the shape and that's all. Nobody wants to know how it's made or why it runs, or even if it runs."

"It used to be that a man would put his thumb through his suspenders and look around and kick the tires, and say, "Well, I guess I'll get that... Then, it got to the point where the woman came in and said she liked the color, she liked the fabric, and she liked the steering wheel, she liked to sit in it. She became a factor, and then the salesman glued himself to the woman. She was very instrumental. The car is fundamental to women." GEORGE WALKER, FORD STYLIST

Billed as *"The All-New Hit of the Highway"*, the exciting new Gregorie-designed '49 Mercurys were ready for public introduction a full two months before the Fords. One of the earliest cars delivered was this *Six-passenger Coupe*, pictured with actress Terry Moore at her home in Palm Springs. Miss Moore starred in numerous movies, including 1949's *Mighty Joe Young*, but her big role was the real-life Mrs. Howard Hughes, becoming one of his few heirs.

Right: The beautiful new 1949 Mercury *Six-passenger Convertible*, in eye-catching *Bermuda Cream*. The price was $2,375, plus $34 for the factory option package that included wheel rings and grille guard. The 110 hp V-8 came standard. Other models in the Mercury line this year were the *Sport Sedan*, *Six-passenger Coupe* (above) and *Station Wagon*.

Left: For those who wanted a little extra style in their lives and could afford to step up to the middle-price bracket the new '49 Mercury was just the car. A salesman at a Tacoma, Washington dealership shows the roomy luggage space of a *Sport Sedan*, offered in optional two-tone paint. Considerate of the farm family, Ford design criteria during the '50s was that all Ford and Mercury cars had to be able to carry two milk cans upright under the deck lid.

Above: No wonder the wide, road-huggin', custom-lookin' '49 Mercs became so popular with American youth. A neat *Sport Sedan*, with an accessory spot light already tilted to the right angle, gets checked out at Rose Bowl Motors in Pasadena, California in the fall of 1948.

Left: Ready for the open road a big, powerful, smooth-riding '49 Lincoln Cosmopolitan *Sport Sedan* awaits its passengers. Buyers had a choice this year of the smaller 121-inch wheelbase "Lincoln", or the 125-inch "Cosmopolitan".

Top left: One of the luxurious new '49 Lincoln Cosmopolitan sport sedans, at $3,125 F.O.B. Detroit, attracts some serious lookers during an evening open house at Rose Bowl Motors in Pasadena, California. It was a time when people dressed for the occasion. Factory installed extras available on this year's Cosmopolitan models included Overdrive, fender skirts, and road lights.

Above: The stately new '49 model lived up to the Lincoln's longtime reputation as the favorite car of U.S. Presidents. Harry Truman visits Des Moines, Iowa in his Cosmopolitan *Six-passenger Convertible* parade car on Labor Day, 1949.

1950 A CAR MAKER'S DREAM

FOR American auto makers and their dealers there was nothing quite like the boom of 1950. The only limit to how many cars could be sold that year was — how many could be built.

"It was unbelievable," recalls Souren Keoleian who was a 19 year-old working on the Ford assembly line in Dearborn. "We had to work six days a week — 9 to 10 hours a day... running two shifts and we couldn't keep up."

Souren, who rose through the ranks at Ford to become head of the service publications department, still remembers clearly the details of that first job with the company: "... My responsibility was to put the taillights on the 1950 Fords... We had the parts by the line and the first thing I did was put a weatherproof sealer on... then a rubber gasket. Then I'd put on the tail light and from the inside attach and tighten two screws with an automatic air gun... Then I'd jump over to the right side and do exactly the same thing. The number of units at that time was 66 cars per hour... one a minute."

Dearborn was just one of many Ford assembly plants across the nation working at capacity trying to fill the unprecedented demand, which was caused by a sudden new prosperity sweeping the nation. People were working and saving money and a lot of then were flocking to the new suburbs in search of the dream of owning their own home with a green lawn, a white picket fence, and a new car parked out front.

For many, the beautiful new '50 Fords they saw advertised on their new TV sets fit that dream — and the rush was on — for a car that looked little different than last year's model.

"... A tremendous effort was made to change the 1950 front-end of the Ford car from a spinner type to some other kind of front-end", recalls styling consultant George Walker's right-hand man, Joe Oros, in his memoirs. "Finally, Ford management bought only a minor facelift which was to change the parking lights from being integrated into the center bar, to a little wrap-around pad detailing with the parking light built in... a new hood ornament... and the lock detailing on the deck lid..."

It was another savvy business move by Company president Henry Ford II. Rather than risk abandoning such a great theme as George Walker's hot-selling '49 Ford, he had decided to stay with the design another year and bank the profits made by barely altering the detail.

The result was a car, with just the right finished look, that helped create the postwar image that a Ford was a jaunty, fun automobile to own and drive.

To capitalize further on this new-found image the company swiftly launched its first big promotion of the convertible — and in mid-year, to counter the new '50 Chevy Bel Aire hardtop — introduced the sporty new *"Crestliner"* in the Ford line, and the rakish new Mercury *Monterey*.

Given little notice was the drive to get weight and costs down on the new '50 Fords by integrating more parts made from high-quality, durable plastic.

Heading this effort was one of the great names in automotive design, Gordon Buehrig, who styled three famous classic U.S. cars — the Duesenberg J, the boat-tailed Auburn Speedster and the Cord 810 — before joining Ford in 1949 as a stylist on the '50 models.

As a result of his work, high impact molded acrylic lenses were crafted to take the place of glass in the tail, parking and back-up lights of the 1950 Fords while the car's window regulators and instrument knobs were now made of butyrate. Molded acrylic had also made its appearance in the classy new three-dimensional steering wheel medallion, and "Ford crest" hood and deck lid ornaments. Small, delicate touches but something Ford designers would have as a new standard to build on for the future. ◆

Left: A booming economy in 1950 saw an unprecedented demand for cars. The One-Millionth Ford passenger car built that year, a convertible, comes off the Kansas City assembly line October 17, 1950. It was given to Oklahoma A & M beauty queen Mary Ellen Ash, pictured with the plant manager.

THE NEW 1950 FORDS

A car for the young-at-heart the new 1950 Ford Custom Deluxe *Convertible Coupe* came in a choice of ten colors, including *Sportsman's Green*, shown here, with *Black, Tan,* or *Green* top fabric to match.

George Walker's '49 Ford was such a huge success that company stylists did little more than change the parking lights on the '50 model for fear of losing the sales magic. But they did make small quality improvements to the car such as easier operating handbrake and new front door courtesy lights.

Left: Some 1950 Fords, including a station wagon, sedans, and convertibles, await the crowds during district dealer introductions of the new cars at St. Paul, Minnesota.

The 1950 Ford line of passenger cars with scores of mechanical improvements, was made public today by the Ford Division of Ford Motor Company.

The new Fords will be on display in the showrooms of 6,400 Ford dealers throughout the nation starting tomorrow.

The '50 Fords will be offered in two lines — the *Custom Deluxe* and the lower-priced *Deluxe* series. A full range of body styles will be available, including station wagons and convertibles. Ford will continue to offer the choice of two engines — the 100-horsepower V-8 and the 95-horsepower Six.

Styling of the 1950 cars includes several distinctive features to enrich the appearance. A colorful new crest, derived from an authentic coat of arms dating back to 17th century England appears on the front of the hood and center of the trunk lid. This is the first crest that has ever appeared on Ford cars. Other styling features include a new hood ornament, re-styled parking lights in new positions and a new ornamental deck lid handle.

Colors are offered in a brilliant new array of choices. There are new colors of broadcloth and mohair for interior finishings. FORD NEWS BUREAU, DEARBORN, MICHIGAN, NOVEMBER 17, 1949

A '50 Ford V-8 chassis moving along the Dearborn assembly line gets brake drum adjustment. Changes this year included a new engine color. *Right:* Installing gauges in a '50 Ford instrument panel in August, 1950.

Left: Atlanta Ford assembly plant workers mate a '50 Ford convertible body to a chassis Dec. 23, 1949. The Atlanta plant enjoyed top honors for consistent quality.

"In the fifties it was pretty well known inside the Company that if you got a car built at Atlanta... that was considered one of the finest cars Ford built because at that time the plant manager was a master of quality..." SOUREN KEOLEIAN, DEARBORN ASSEMBLY STAFF

Amid patriotic bunting and snappy *"Come See It! Come Drive It!"* banners, an exhibit of new 1950 Atlanta-built Fords are ready for a district dealer preview at the Atlanta, Georgia City Auditorium, November 7, 1949. In the foreground is a nifty *Sportsman's Green* convertible. The Ford crest ornament on the deck lid was a new styling touch.

Right: An evening crowd of mostly men looks over the new '50 Fords at the showroom of Les Kelley, the Los Angeles dealer who founded the famous "Kelley Blue Book" car price guide. These cars were from the Long Beach assembly plant where Ford had struck so much oil its pumping would eventually sink the place.

1950 CONVERTIBLES

The smart new 1950 Ford Convertibles now on display in dealers' showrooms offer a choice of eleven new exterior paint colors, including four metallic finishes.

Two of the new colors are exclusive in the Convertible line. They are *Sportsman's Green* – a beautiful chartreuse which comes with a *chartreuse* and *black* leather trim or a combination of *black* leather and *Bedford Cord*, and *Matador Red Metallic* – a medium red color trimmed with *black* and *red* leather, or with the *Bedford Cord* combination.

In addition to the new colors, the 1950 Ford convertibles feature an improved top mechanism, a non-sag front spring, new sponge rubber front seat cushions which provide added comfort and a longer, improved regulator handle permitting the raising and lowering of the quarter windows with great ease.

Other colors in the 1950 Convertible line are: *Dover (light) Gray, Palisade (light) Green, Hawthorne (dark) Green Metallic, Osage (grayish) Green Metallic, Sheridan (dark) Blue, Cambridge Maroon Metallic, Black,* and *Sunland Beige.*

Like all the 1950 Fords, the new convertibles are powered with the new, improved V-8 and 6-cylinder engines which are notable for their smoother flow of power and quieter operation. Additional insulation and sound deadener in doors, panels and floor adds to the silent operation of the new convertibles. FORD NEWS BUREAU, DEARBORN, NOVEMBER 27, 1949

Ford's big sales push in 1950 was the new Ford convertible. More than 50,000 were sold that year, compared to Chevy's 32,800, to make it America's favorite soft-top. Even Hollywood got in on the act. Gateway Motors of Albany, New York provided this new '50 Ford *Convertible* for screen hero Burt Lancaster who was in the city promoting his latest film, *The Flame and the Arrow*. The car's spiffy extras include a continental spare and rocket hood ornament.

Right: The 1950 "Miss Maid of Cotton", Elizabeth McGee, from South Carolina, gets the loan of a continental kit-equipped new '50 Ford *Convertible* from Burl Berry Ford for a tour of San Francisco.

Dan Aschcraft, of Beverly Hills Ford in California, shows the classy continental rear tire kit his agency installed on a new '50 Ford convertible. The hot accessory had suddenly become the rage, especially on the West Coast, and was a high-profit dealer add-on. For protection from rear-enders, the kit required installation of heavy-duty bumper guards. Other extras on this car are fender skirts and back-up lights.

THE NEW CRESTLINER

The *Crestliner* — a distinctive new two-tone sports sedan with black basket weave vinyl covering the steel top and a striking new airfoil custom paint panel on either side — has just been introduced by the Ford Division of the Ford Motor Company.

The new model, only one of its kind in the field, is available in *Sportsman's Green* with *Black* top and *Black* airfoil side panels, or in *Coronation Red* with *Black* top and side panels. Interior trim of the new models resembles European-type sports cars.

The *Crestliner* achieves its distinctive external appearance through the use of an airfoil panel of jet black which covers most of the front fender, three quarters of the door and extends well into the rear side panel. It is outlined with a narrow border of stainless steel molding. The word "Crestliner" in a gold-colored die cast appears on the side of each front fender.

Completely new full wheel-discs of chrome with circular depressions painted black are standard equipment on the new model, as are the fender skirts and two side view mirrors. A protective and decorative molding of polished stainless steel extends along the bottom edge of the car.

Inside, the *Crestliner* has a two-tone instrument panel with the body color at the top of the panel and the remainder painted a satin-finish black. The top half of the door panels are of body color vertical *Bedford Cord* and the lower portion of leather-grained material with rows of horizontal stitches along the bottom quarter.

Seat backs and cushions are covered with *Bedford Cord* material in pin-striped pattern to match the car's exterior color. The sides and the top facing of seat backs are trimmed with black deep-buffed *Leather*. The back of the front seat is black to contrast with the car's exterior, and the headlining is black grained art leather. FORD NEWS BUREAU, ROTUNDA, DEARBORN, JULY 9, 1950

The first assembled Ford *Crestliner* is pictured outside the Dearborn Assembly Plant, June 22, 1950. Introduced to the public three weeks later, the distinctive new sport model, with "airfoil" side panels and fabric-over-steel top, was designed to spark summer sales. Ford customers responded by snapping up over 17,000 of them.

Right: A new Ford *Crestliner* on exhibit at the exclusive Red Run Country Club in Royal Oak, Michigan, June 30, 1950.

"The original model of the *Crestliner* is attributable to Bob McGuire's efforts as the chief stylist for the Ford studio. It proved very succesful in the marketplace . . ." JOE OROS, FORD STYLIST

Winning an Oscar for her role in the big 1948 screen hit, *Johnny Belinda*, was the highlight of actress Jane Wyman's 36 films, dating back to 1937. She is pictured at her home in Hollywood with one of the rewards, a new '50 Mercury *Convertible*. At the time she was married to actor and future U.S. President, Ronald Reagan. Their other car was a new Lincoln coupe.

Left: In mid-1950, Mercury dealers got their own version of the Ford *Crestliner* in the form of the new vinyl-topped *Monterey* coupe.

1950 PRESIDENTIAL LINCOLN

The most famous White House limousine was President Harry S. Truman's stretched and specially-equipped 1950 Lincoln Cosmopolitan. He is pictured with the car while on a visit to Detroit City Hall, July 28, 1951. In 1954, President Eisenhower had Ford fit it with a clear plastic top so he could see, and be seen better and from then on it would be known as the "Bubbletop", serving until 1961.

A 7-passenger Lincoln Cosmopolitan Convertible, the last of ten special 145-inch wheelbase Lincolns constructed by the Lincoln-Mercury Division for the White House, has been delivered to President Truman in Washington.

The long-low custom-built convertible is painted black and has white sidewall tires. It has cherry-red and black genuine leather upholstery, gold-plated interior appointments, a tan top and two comfortable folding seats. An unusual feature of the car is the use of chrome fender side moldings on the rear as well as the front fenders.

The car is equipped with special disappearing steps on each side of the rear fenders for the Secret Service men who guard the President. Special chrome hand grips on the rear quarter panels serve as supports for the President's guards.

The President's convertible, as well as the nine special custom-built Lincoln Cosmopolitan limousines which have been delivered to the White House in recent weeks, are all leased to the Government under a contract with Ford Motor Company.

Each car is powered by a regular high-compression 152 horsepower V-type Lincoln 8-cylinder engine, and each is equipped with heavy-duty hydra-matic transmissions.

Specifications include a warning siren and special flashing red lights that replace the car's conventional road lights. LINCOLN-MERCURY NEWS BUREAU, DETROIT, JUNE 12, 1950

1950 MERCURY PACE CAR

In line with established Speedway policy, the pace car for the 1950 International 500-Mile Race at Indianapolis on May 30 will be a stock Mercury convertible, with engine specifications identical to the car which won the recent 750-mile Grand Canyon run with the remarkable average of 26.52 miles per gallon of gasoline.

It will be driven by Benson Ford, vice president of Ford Motor Company and general manager of the Lincoln-Mercury division, with Wilbur Shaw as a passenger. Shaw is president and general manager of the famous two-and-a-half mile oval.

The pace car will be presented to the winning driver at the Victory Dinner following the race.

The late Henry Ford spent many hours in gasoline alley with the outstanding drivers and mechanics of early speedway history, also serving as honorary referee of the 1924 race.

His son Edsel paced the 1932 race in a Lincoln. A Ford V-8 driven by Harry Mack paced the 1935 Indianapolis classic, and Henry Ford II led the 1946 field across the starting line in a Lincoln Continental when racing was resumed after the last war. FORD ROUGE NEWS, APRIL 7, 1950

For the first time ever, Mercury was selected in 1950 as the official car of the Indy "500" race. Three of the factory-prepared cars are pictured on their way to Indianapolis. The two sedans were for press and radio use and the convertible was for track official Tony Hulman.

Right: Benson Ford in the official 1950 Indy "500" Pace Car, a new Mercury *Convertible*, chats with TV show host Ed Sullivan. Benson was on a roll. He had just signed Lincoln-Mercury to sponsor the "Ed Sullivan Show" and got his Mercurys named official cars of the race classic. The pace car went to the race winner, Johnnie Parsons.

Ford built just 1,343 of these "windshield" model F-1 trucks in 1950. A big part of the production went to a custom fleet order from the Good Humor ice cream company for use in New York state. In a company publicity shot one of the milk-white Ford-built jobs, slated for delivery to Coney Island, is pictured with the famous bell-ringing "Good Humor Man" in white uniform and change belt dispensing his curb service.

Right: When you're loading up the car every dog knows it's time for a ride! A well-off young Michigan couple prepares to take off for the weekend in their brand-new 1950 Mercury *Station Wagon*. Just the thing for upscale two-car suburbanites, it seated eight passengers in 3-2-3 order or you could remove the center and rear seats to haul bigger things like camping gear and canoes. Traditional wood side panels over the steel underbody were in maple and mahogany and you could get your wagon with either *Tan, Red,* or *Green* genuine leather upholstery.

"... There's a mysterious, indefinable something ... to good sales appeal ..."
GEORGE WALKER, FORD STYLIST

1951 AERO-STYLED BEAUTY

BY the time the new '51 Fords were introduced at showrooms in late November, 1950, the national car-buying frenzy that started earlier in the year had kicked into second gear. This time because of a place called Korea.

A few months earlier, North Korea had invaded South Korea, drawing the U.S. into the conflict and setting off a scare there would be shortages like those in World War II. The rush to buy a car — any car — while you still could, created a new wave of demand. Then it got worse for both buyers and makers as auto production began slowing as a result of steel shortages caused by the new war economy.

Lee Kollins was fresh out of the U.S. Marines in 1949 when he got a job in Ford Public Relations. He remembers how worried the company was... "Mr. Ford (Henry II) had to go to Washington to plead for an increased steel allotment. He even wanted Kaiser's allotment. Kaiser (the auto maker) was floundering at that point."

It was this uneasiness about the future, and war jitters in general, that brought an estimated 6-million people, during the worst winter weather in memory, jamming in to see the new '51 Fords when they were first introduced in November, 1950. Dealers hadn't seen anything like it since the crowds of 1946-47, when the pent up demand for cars and short supply saw deals being made under the table. People were signing up on waiting lists—anything to get one of the new '51 Fords that were coming so slow off the assembly lines.

Once again the stampede for the new Ford overshadowed its handsome styling — even with the car's head-turning "double-spinner" front-end treatment that gave it a unique look all to its own.

"The '51 Ford was an extension of the '49 Ford", says Joe Oros, from George Walker Design, who helped style it. "The '49 being so highly successful and the '50 equally, we did not want to abandon the theme... So, we continued with the spinner front-end as dual spinners... We literally added two pods on the horizontal grille bar... in keeping with the aircraft theme of the twin engine."

It was an aero-styled beauty in an unreal market but even then Ford was looking for ways to capture more sales from the competiton by making its cars more appealing to women buyers. Offered for the first time was an optional "Fordomatic Drive" automatic transmission that made driving easier, and in March, the sporty new *Victoria* "hardtop" model was introduced.

Ford wanted to bring out its first "hardtop convertible" with the all-new '52 models but the dealers were demanding something sooner to compete head-on with the Chevrolet Bel Aire. Gordon Buehrig was working in the Ford styling studios at the time.... "John Oswald (then head of Ford styling) came in one day and said, 'Can you take a convertible — '51 convertible — and make it a hardtop?' Well, of course you already had the windows for the door and the quarter window, so it's just a matter of doing the new top for it. We did it with a band corrugated of metal up over the rear window which gave it a styling distinction.

"We had a quarter-scale clay model that someone else had built of the convertible, and we brought it in to our studio and quickly transformed it into this model. It sat there for several days, and one day I came in, and it was gone. I inquired about it and found out that it had been sent out to the job shop where they were engineering it and tooling it for production... and I later found out that George Walker, who was a consultant at that time, claimed credit for it..."

Not one to miss a good opportunity to impress Henry Ford II with a good idea, even if it wasn't his own, the irrepressible George Walker had scored again! ◆

Left: A 1951 Ford Deluxe *Business Coupe*, posed against a Rouge Plant backdrop for a magazine cover, makes an impressive statement of American war-time industrial power.

The action stops for the publicity camera as freshly built '51 Fords (a *Convertible*, *Victoria* and *Tudor*) are loaded aboard a 1950 Ford F-5 transport at the Dearborn haul-away lot. They are bound for an outlying dealer where the three cars will go in the showroom for a combined price of about $5,100.

Left: Built with pride, a '51 Ford Custom *Convertible Coupe* in popular *Coral Flame Red* reaches the end of the line at Dearborn Assembly. From here, all the soft tops went on to another line *(right)* for hand-fitting one of three choices of tops and eight choices of leather interior combinations to the car's body color. Only Ford's top craftsmen did this work and many of them dated back to the Model A.

"Let's Send Everybody Our Best!" Sign over the Dearborn assembly line, 1951.

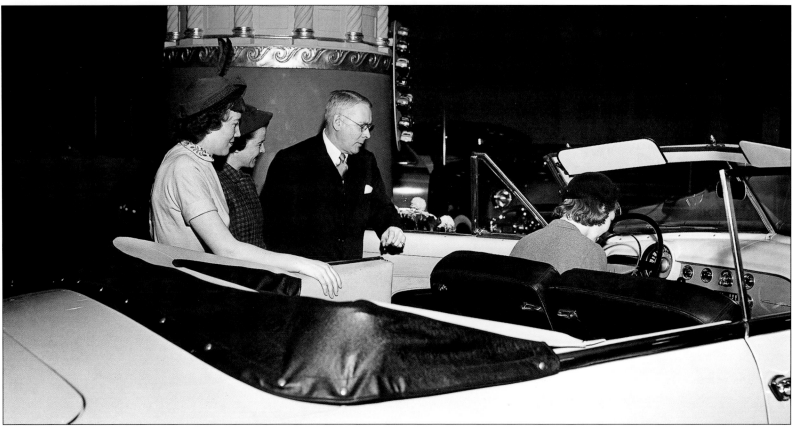

1951 FORDS DEBUT

Dealer's and the nation's press get their first look at the new '51 Ford models at a special preview at Detroit's Masonic Temple during the week of November 10, 1950.

Left: Checking out the dreamy new '51 Ford *Convertible* at the Detroit Dealer Preview. *Top left:* Getting an up-close look at the sporty new *Victoria* model a full four months before its introduction to the public. New Ford features this year included an extension treatment of the rear fenders, and three-lens taillights.

"Beauty is what sells the American car. And the person we are designing it for is the American woman." GEORGE WALKER, FORD STYLIST

The 1951 Fords, featuring Fordomatic Drive, the new automatic transmission, will be introduced in Ford dealerships throughout the nation Friday, November 2nd.

Introduction of Fordomatic climaxes several years of intensive engineering research by the company to develop a fully automatic transmission that is simple to operate, economical, long wearing and easy to service. The device, which will be optional at extra cost, combines the advantages of a hydraulic torque converter and a planetary gear train to provide maximum performance and a smooth and economical flow of power.

A lowered hood, a new dual-spinner grille and a wider, longer wrap-around bumper give a more massive appearance to the front end. Parking lights are restyled and larger chrome headlamp rims extend beyond the lenses. Added chrome and new, larger tail lamps add to the appearance of the rear end.

The Ford Deluxe line includes the *Business Coupe, Tudor* and *Fordor* models. In the Custom line are the *Club Coupe, Tudor, Fordor* and *Convertible*. The *Country Squire* Station Wagon and the *Crestliner* complete the passenger car line. The V-8 engine or the Six are optional on all models except the *Convertible* and *Crestliner*.

New colors are *Coral Flame Red, Alpine Blue, Culver Blue, Sea Island Green, Greenbriar Metallic,* and *Mexicalli Maroon*. Other colors are *Sheridan Blue, Silverstone Gray, Hawaiian Bronze, Hawthorne Green* and *Raven Black. Sportsman's Green* is reserved for the *Convertible* and *Crestliner. Coral Flame Red* is reserved for the *Convertible*. FORD NEWS BUREAU, DEARBORN, NOVEMBER 23, 1950

NEW FORD VICTORIA

The new Ford *Victoria* captures the sports lines and beauty of the convertibles, yet offers the all-weather comforts and advantages of a sedan for summer driving and winter service.

Ford's newest offering in its 1951 line of passenger cars features an all-steel top and nearly 3,000 square-inches of exposed glass area for clear visibility, including a huge rear window and roll-down side and rear-quarter windows.

The *Victoria* is an all-purpose touring car. It is handsomely trimmed and richly appointed both inside and out, with ten exterior finishes and harmonizing interior trim combinations.

Seats are upholstered in durable Ford "Craftcord" fabrics with genuine deep-buffed leather and vinyls. Carpeting, headlining, instrument panel and garnish moldings are in blending colors. There are five single-tone and five two-tone body colors and three interior color ensembles to harmonize.

The Ford crest and a gold-finished *Victoria* name plate adorn the "Safety Glow" instrument panel which distinguishes 1951 passenger models in the Ford line. Bright metal trim frames the rear window.

Single-tone body colors and harmonizing trim includes *Mexicalli Maroon Metallic* exterior with brown interior ensemble; *Raven Black* or *Alpine Blue* exterior with gray ensemble; and *Hawthorne Green Metallic* or *Sea Island Green* with green ensemble.

Two-tone colors are *Hawaiian Bronze Metallic* body and *Sandpiper Tan* top with brown ensemble; *Alpine Blue* body and *Silvertone Gray* top with gray interior; *Greenbrier Metallic* body and *Sea Island Green* top or *Sea Island Green* body and *Raven Black* top, or *Sportsman's Green* body and *Raven Black* top with green interior ensemble. FORD NEWS BUREAU, DEARBORN, MARCH 16, 1951

"It's smart as a convertible! Snug as a sedan!"

Ford's first in a long-line of "hardtop convertibles" was the hot-selling new '51 Ford *Victoria* aimed at the sporting crowd to compete with the Chevy Bel Aire. Buyers now wanted more convenience, along with good looks in their cars. Twenty-five percent of all 1951 Fords were sold with the new Fordomatic transmission.

Right: **Pictured at a Detroit dealer, Miss June Lockwood shows how the "pillarless" doors of her new '51 Ford *Victoria* give the car a clean, open-air convertible look from the windshield post back. Ford's *"safety of a steel top"* advertising slogan was a big selling point with the ladies.**

Left: A Ford staff photographer took this neat publicity shot of a 1951 Ford Custom *Convertible Club Coupe* at the coal docks of Ford's River Rouge plant in Dearborn. The car was fresh off the plant's assembly line in convertible-only *Sportsman's Green* with optional black top. Deck hands of the *E. G. Mathiott* look on. Ford had its own fleet of boats bringing raw materials and iron ore to its huge steel-making operations at the Rouge.

Above: A *Sportsman's Green* '51 Ford Custom *Convertible Club Coupe* in one of two *Chartreuse* and *Black* Leather interior combinations that could be ordered with that body color. The new *"Air Wing"* steering wheel, *"Semaphore Drive"* selector, and *"Safety-Glow"* control panel with a glowing ring illuminating the speed numbers, showed the car's airplane-styling influence. All the Ford models now had "turn-key" starting.

"Long looking, low looking, and good looking..." 1951 FORD ADVERTISING LINE

Above: The Ford brothers take time for a portrait in 1950 at their styling studios with a new '51 Mercury *Sport Sedan*. Left to right are Benson, 31, Henry Ford II, 33, and William Clay, 25. Firmly in charge of the company they had inherited, at this time Henry II was President, Benson was Vice-President of Lincoln-Mercury, and William was a company Director. Their youth is shown in all the great-looking cars they were producing.

Top right: Famous to a nation of car fans as an automotive authority and reporter for *Mechanics Illustrated* magazine, Tom McCahill tries out a new '51 Mercury *Convertible*.

Right: The new '51 Mercury Six Passenger *Sport Coupe* in *Tomah Ivory*. Two new options to go with the 112 hp V-8 were a Borg-Warner *"Merc-O-Matic Drive"*, or *"Touch-O-Matic"* Overdrive. American cars were looking bigger. The '51 Mercurys were part of that trend with a big wraparound grille and rear fenders that swept beyond the deck lid.

52

NEW COUNTRY SQUIRE

The "Country Squire", a new "double-duty" station wagon which can be converted from a comfortable eight-passenger carrier to a one-level-floor cargo wagon in three minutes, has been introduced by the Ford Division of the Ford Motor Company.

The "Country Squire" features new mahogany-grained steel side panels, framed with natural maple or birch. The new steel side panels add longer life and greater strength to the body.

An all-steel rear deck on which the spare tire is mounted is also a new feature of this model. By placing the tire outside, Ford engineers have provided for its removal quickly and easily without disturbing the inside cargo.

Designed as an all-purpose carrier, the new model fulfills the need for a "Sunday calling" station wagon which can be changed quickly to a "Monday hauling" load carrier with a minimum of effort.

To convert to maximum floor area, the cushion of the two passenger stowaway seat is pulled forward and the bottom of this seat, covered with heavy ribbed composition material, becomes part of the level-loading floor. The back of the stowaway seat then folds forward to add to the level floor surface.

Ford's automatic overdrive is available on the new model at extra cost. The new station wagon is available in the Ford Custom Deluxe line of body colors. FORD NEWS BUREAU, ROTUNDA, DEARBORN, JUNE 30, 1950

A handsome 1951 Ford *Country Squire* Station Wagon, last of its kind since 1929 to have a famous hand-crafted body from the company's Iron Mountain plant in the hardwood forests of Michigan's Upper Peninsula, is pictured at the Rouge plant in Dearborn. Plymouth and Chevrolet, with their cheaper easier to maintain all-steel wagons won the most sales in 1951, forcing Ford to offer its own steel-bodied models the following year.

Right: Under the tent at the annual Bloomfield Hills Hunt Club horse show in Bloomfield, Michigan, an exhibit of new '51 Fords appropriately includes a genteel *Country Squire* Station Wagon at $2,021 Detroit, and a sporty *Convertible Coupe* priced at $1,867.

Left: Among the rarest of all post-war Lincolns was the superb 1951 Cosmopolitan *Six-passenger Convertible*. Incredibly, just 857 were sold. Meanwhile, one of these luxury soft tops finds some admirers at Ford's tent exhibit at the 1951 Bloomfield Hills Hunt Club horse show. The car's price was $3,700 plus such extras as Hydra-Matic and hydraulically-operated windows and seats.

Above: Grand Prize winner of the 1951 840-mile Los Angeles to Grand Canyon Mobilgas Economy Run was this '51 Lincoln Cosmopolitan sedan entered by Bob Estes, the Ford dealer in Inglewood, California. It averaged 66.484 ton-miles per gallon. At the end of the rugged run through deserts and over mountains, it gets a reenactment of its start by an official at the Grand Canyon's rim.

1952 THE NEW CUSTOM LOOK

DURING the 1950's it took Detroit about three years to design a car from the germ of an idea to actual production. This means that sketches of what the new '52 Ford might look like three years down the road were going on the drawing boards about the same time the '49 models were in the showrooms.

It was an awesome planning job in the days before computers, that started with thousands of drawings by the Styling Department. First came the idea sketches, wilder than what was really possible. When all the workable ideas and angles were agreed on, they were blocked in full scale on a huge sheet of paper. Then a three-eighths model of the full-size car was made, and changed many times. Finally a full scale clay model was sculptured, complete down to the last chrome molding.

One man whose styling would influence details of the all-new '52 Ford was Gordon Buehrig:

"They put me in charge of the body development studio. It was (here) that the models other than the sedan were to be developed. The first project we had was on the '52 Ford on which the sedan had been approved . . . it was our job to design the convertible and the station wagon using maximum inter-changeability with the sheetmetal from the sedan . . ."

Buehrig and others worked under a chief stylist such as Tom Hibbard or Frank Hershey, depending on who was in charge. Then the man with the golden touch, George Walker, would walk in and give his opinion as the company's outside design consultant. Naturally, the Ford designers resented this arrangement, but went along with it.

Walker helped refine the design, shaving an eighth of an inch here, changing a curve there. At his George W. Walker studios in Detroit he would direct his assistants to give it something of their own. Then Walker's incredibly accurate artistic sense of what would sell three years into the future when the final product was built, came into play. Finally, he would wrap the best of all these ideas into a "package" and "sell" it to his friend Henry Ford II in one of his legendary P.T. Barnum-style presentations.

What the young Company president went for in the '52 model was a not-too-wild, slick-but-conservative, new Ford with a lot of Walker touches. It had "frenched" headlights, a new twist to the airplane theme with a "triple-spinner" grille, a one-piece windshield and backlight to give the car a wider appearance, and a new rear-end theme that would become a Ford hallmark — round tail lights, designed by Joe Oros after the jet exhausts on aircraft.

Nice as they were, the '52 Fords were not big sellers, primarily because of the ongoing effects of the Korean War and an introduction two months late that caused a shortened production year. As a result, in terms of total sales, it would be one of Ford's worst years ever. ◆

Right: A picnic in the park with an attractive new '52 Ford *Sunliner* Convertible in *Sungate Ivory*. Its pleasing lines show the work of famed auto designer Gordon Buehrig.

THE 1952 FORDS

"Frenched" headlights and jet taillights gave the completely restyled new 1952 Fords a streamlined, longer appearance as seen in this pretty *Sunliner* convertible in the new top-of-the-line Crestline series. Push-button door handles and the gas cap centrally located behind the swing down center rear license plate were new features this year.

Left: Prominently displayed under the Ford Motor Company tent at the 1952 Bloomfield Hills Hunt Club horse show, against a backdrop of other models, a Crestline *Sunliner* convertible awaits the crowds. Ford sales were off this year, primarily because of production shortages caused by the ongoing war in Korea.

Ford Division of Ford Motor Company offers three completely new lines of Ford passenger cars for 1952.

There are a total of 18 models in the three new lines — eight in the Mainline series, seven in the Customline series, and three in the Crestline series. The new Ford's styling identity is preserved in the low, wide lines of the hood and front-end of the new cars, centering around a triple-spinner, air-scoop grille arrangement. The new cars have curved one-piece windshields and rear windows, and their body lines sweep backward from extended headlights along the hood and higher fender line to the jet-tube taillight sections.

Among the 1952 models are three completely new utility passenger vehicles, one for each of the three lines of cars. The *Ranch Wagon* in the Mainline series is a two-door, six-passenger unit with all-steel body. In the Customline there is the *Country Sedan*, a four-door eight-passenger vehicle with all-steel body. In the Crestline is the *Country Squire*, another four-door eight passenger model, with wood trim over the steel side panels.

A range of 12 two-tone exterior body color combinations is available for closed car models and there are 12 single body colors, all with matching and harmonizing trim and upholstery.

Ford Mainline body styles for 1952 are the *Business Coupe, Tudor, Fordor* and the *Ranch Wagon*. Customline units are the *Tudor, Fordor,* and the new *Ranch Wagon*. Customline units are the *Tudor, Fordor, Club Coupe,* and the all-metal *Country Sedan* Station Wagon. The Crestline cars are the *Sunliner, Victoria* and *Country Squire*. FORD NEWS BUREAU, DEARBORN, MAY 20, 1952

A demonstrator at the 1952 Michigan State Fair explains the features of the new '52 Ford, cutaway to show details. The car's venerable V-8 engine was stepped up to 110 hp and the optional overhead-valve "Six" was upped to 101 horsepower. New Ford "firsts" on the '52 models included a brake master cylinder in an easier accessed location on the firewall, and suspended brake and clutch pedals so *"there are no shafts to stub your toe or floorboard holes for winter winds to whistle through."*

Left: The new Ford *Victoria* model at the 1952 Bloomfield Hills Hunt Club horse show. More glass area all around and the new one-piece curved windshield gave the '52 Fords a wider look.

Whitewall tires on a 50's Ford made a big difference between a plain-looking car and a youthful, exciting one. Proud of his brand new '52 Ford *Victoria*, this Salt Lake City, Utah gent poses in his Sunday's best for a publicity shot at his local tire store where he traded in the factory blackwalls for a niftier set with better treads for snow country.

Assembly action at the Atlanta plant on a summer day in 1952. In a reversal of regular Ford procedures the front-end here is being decked onto a V-8 chassis, before the body. Frame numbers were stamped in with a power clamping device earlier on the line.

Left: A red *Pickup* cab rides on the line with a string of freshly enameled '52 Ford sedan bodies at Atlanta, where every fourth vehicle built was a truck. At this point the body got its hood and wiring.

CODE MARKINGS

Have you ever wondered what the code markings stamped on various parts of your car mean?

These represent another means by which Quality Control constantly strives to improve Ford Motor Company products. The code markings are stamped on parts to identify them as to time of inspection. In engine assembly, for example, four symbols are stamped into each block at the time of passing final inspection.

The first symbol represents the year; the second, the month; the third, the day of the month; and the fourth, the individual inspector who okayed the assembly. The combination of symbols to be used varies with the part being stamped. For example, it may be sufficient to identify one part only as to year and month; another part may require year, month and day; and in other cases the shift and hour may also be included.

This system enables Quality Control troubleshooters and service representatives to refer back to a specific time of inspection in the event of any failure of the part later. FORD ROUGE NEWS, OCTOBER 17, 1952

Above: Gordon Buehrig's Ford design team fashioned three new steel-bodied station wagons for 1952: the no-frills *Ranch Wagon*, the fancy *Country Squire*, and this modest *Country Sedan* for the middle-income bracket.

"... They wanted a new line of station wagons... Chevrolet had come out with an imitation wooden body made out of steel. So, McPherson (Ford Chief Engineer) told me that's what he wanted to do..." GORDON BUEHRIG, FORD STYLIST

Right: The classy new '52 Mercury Monterey *Convertible* was listed at $2,390 F.O.B. Detroit. Fender skirts came standard on Montereys. Options included Merc-O-Matic, Overdrive, electric windows and seats, curb buffers, custom steering wheel, and tinted glass. Mercurys this year, like the Fords, got new suspended pedals, center gas-filler, and one-piece windshield.

67

1953   *FORD AT FIFTY*

FORD Motor Company had many good reasons to believe that 1953 was going to be a banner year. The war in Korea that had been so costly to business was all but over. The future was looking bright. And it had just launched the biggest celebration in its history to mark the fiftieth anniversary since its founding by Henry Ford in 1903.

And what a party! Two years in the making, the jubilee was on a grand scale like nothing seen in the auto industry before or since!

Among the commemorative souvenirs designed for the event were the cars themselves. As they rolled into the dealers' showrooms specially decorated for the gala introductions, each had some kind of anniversary emblem to make them unique. The beautifully-styled new Fords had a gold, red and blue medallion on the steering wheel; the sleek new Mercurys had a special gold crest dash emblem, and the elegant Lincolns had a gold dash medallion, plus a gold-plated hood ornament and body side moldings. Even the Ford trucks and tractors had emblems to designate them as anniversary models.

The mementos poured into the Ford showrooms to be handed out to anyone who came in to take a look at the new cars: two and a half million anniversary calendars, featuring a series of Ford history illustrations by famed American artist Norman Rockwell; countless commemorative coins struck from one of Rockwell's illustrations, a portrait of the profiles of Henry Ford, Edsel, and Henry II; toy models of this year's Indianapolis "500" Pace Car, a '53 Ford *Sunliner* Convertible . . . The list went on . . .

In Dearborn, *The American Road,* a major auto industry film narrated by Raymond Massey, was produced to tell the Ford story; the Ford Archives were dedicated at Fair Lane, Henry and Clara's old home; a half-million copies of a complimentary picture book was published, *Ford at Fifty,* in two versions, one for employees and one for customers.

Ford's founding date was June 16th. On national television the night of June 14, 1953, Ed Sullivan and his Lincoln-sponsored TV variety show gave Ford a big salute; the next night Ford made history by sponsoring the first two-hour variety show ever seen on American TV. A celebration of Ford's anniversary seen by millions, it was an extravagant Leland Hayward production the *New York Times* called, "epochal" and "breathtaking". Narrated by Edward R. Murrow and featuring songs and skits from such musical hits as "South Pacific", the show starred Mary Martin and Ethel Merman, with such guests as Bing Crosby, Rudy Vallee and Marian Anderson.

On the Company's birthday the new Engineering and Research center in Dearborn was dedicated, followed the next day by the reopening of the Ford Rotunda visitor's center across from Ford Motor Company headquarters. Before it closed its doors to the public for the duration of World War II in 1942 the gear-shaped Rotunda building, with its encircling "Roads of the World" where visitors took rides in the new Fords, was one of the nation's most outstanding tourist attractions.

Lee Kollins was there as a Ford host when the Rotunda made its nighttime reopening, lit like a giant birthday cake. . . . "We had a special Lincoln on display in front of the auditorium. It was done in gold and pearlescent (the *Maharaja* show car, page 83). Lowell Thomas was there as master of ceremonies . . . What a show!"

And what a year for Ford! By the time it ended, there was little doubt that the company that put the world on wheels was still going strong — more popular than ever. ◆

Left: Getting ready for a milestone year, Benson, William, and Henry Ford II pose with futuristic X-100 (front) and X-500 concept cars for the 1953 cover picture of *Ford at Fifty,* the Company's 50th Anniversary book.

FORD, <u>Worth</u> more when you buy it. <u>Worth</u> more when you sell it!

A large crowd looks over the new '53 Ford *Fiftieth Anniversary* passenger models at Stuart Wilson Ford on Michigan Boulevard in Dearborn. It was a time when new model introductions in the fall of each year were a special community event eagerly anticipated. Predictably, there were always car fans trying novel ways to sneak a peak at the new cars before the official introductory day.

ANNIVERSARY FORDS

A new '53 Crestline *Victoria* in stunning *Sungate Ivory* is displayed at Francis Ford in Portland, Oregon to make the connection between fashionable Fords and fashionable women. The 1953 Fords were designed to please the ladies, with new *"Miracle Ride"* suspension, softer seat cushions, power steering on V-8s, and optional soft-green tinted safety glass.

The 1953 Ford – marking the 50th anniversary of the Ford Motor Company – will go on display in 6,400 Ford dealerships across the country Friday, December 12th.

A massive new grille with a center spinner characteristic of recent Ford design and a low, road-hugging look, advance the modern Ford styling. And an outstanding improvement in suspension, termed "the miracle ride", heads the list of mechanical improvements.

To designate the 1953 Ford as the *50th Anniversary* car, a medallion has been placed on the top of the steering column in combination with a new half-circle horn ring. Around the Ford crest in the center of the emblem are the words *"50th Anniversary – 1903-1953."*

Emphasizing the longer look of the 1953 Ford is a new chrome molding through the center of the rear fenderline and new jet-tube taillights with a larger signal area more easily seen from the side or rear. A decorative chrome deck handle, mounted below the Ford crest, with a concealed, weather protected key-opening and key spring return, has been added to the new model.

Known as the world's largest builder of V-8 engines, Ford features its *Strato-Star* V-8 engine as the only V-8 in the high volume field. Its 110 horsepower is the highest in this field. Also available on the 1953 Ford is the *Mileage Maker Six*.

Additional features offered in the 1953 Ford cars include a new jet-wing hood ornament, and new oblong parking lights, and a new one-piece wrap-around rear window in the Victorias.

Ford continues to feature center-fill fueling with the gasoline cap concealed behind the license plate, easier steering, push-button door handles and rotary latches, and extra large luggage compartment. FORD NEWS BUREAU, DEARBORN, DECEMBER 10, 1952

Ford station wagons had become so popular they were now a common sight on the assembly lines with cars. Here, a '53 *Ranch Wagon* model reaches the end of the line at Dearborn. A new feature of the '53 Fords was a fiber glass insulation blanket on the underside of the hood to sound-proof engine noise.

Left: Installing the glass in a 1953 Ford *Victoria* body on the trim line at the Chicago Assembly Plant September 24, 1953. Among selling features of the '53 passenger models was the new "hull tight" body construction designed to seal out water, dust, and drafts.

Right: Metal finishing Ford bodies at Chicago in 1953 with the traditional hammer and dolly, hot lead, and grinder.

Left: The new '53 Ford Crestline *Sunliner* was available in a dozen colors, including two shades of red: *Carnival Red* or *Coral Flame Red* with harmonizing *Black Leather* and *Red Vinyl* interior. Like the other two models in the '53 Crestline series, the *Victoria* and station wagon models, the convertible frame was reinforced for more strength through the door areas. Dress-up items at extra cost this year included full wheel covers, tinted glass, leather trim, and Coronado Deck "continental kit".

Above: Inspecting under the hood of a '53 Ford V-8 in a time when gas was about 22 cents a gallon and a stop at a service station often got you a smiling uniformed attendant who checked all your car's fluids, washed the glass all around, and aired the tires. Ford engine options this year were the 110 hp "Strato-Star" V-8 and the 101 hp "Mileage Maker" Six.

"It was a showcar on the floor of the Rotunda . . . a tomato-red '53 Ford hardtop with a continental kit . . . it was very sexy . . . it fit my style. When I learned it was going to be replaced I followed it to the company's used car lot . . . This car was spotless and I bought it." LEE KOLLINS, FORD ROTUNDA, PUBLIC RELATIONS

CORONADO DECK

A new Coronado Deck conversion which provides Ford passenger cars with the smart appearance of a continental-type rear wheel mount without extensive and costly alterations, is now available as an accessory at Ford dealerships.

While resembling a "built in" wheel carrier, the new accessory requires no special alterations to the car body and adjacent parts other than relocating the "Fordomatic" or "Overdrive" nameplate. And the spare wheel remains in its regular, easy accessible location.

The cover of the Coronado Deck is one-piece, heavy gauge sheet metal with the Ford crest in the center of a cap which covers the luggage compartment lock. This cover is bolted to the regular deck lid. Easy access to the center-fill tank filler is gained through the special license plate frame which is equipped with a stainless steel handle.

Because of the simplicity of its design and the absence of any body alterations, the new Coronado Deck conversion retails at a fraction of the cost of non-Ford conversions. FORD ROUGE NEWS, DECEMBER 5, 1952

The 1953 Ford pace car during time trials at the famed Indianapolis Speedway. The car's crew, in crisp team uniform takes a break to pose for a company photo. Their job was to make sure the factory-tuned soft top ran right and looked right on race day. Beautifully prepared, the *Sunliner* in white with gold lettering, was equipped with Kelsey-Hayes chrome wire wheels, twin spotlights, fender skirts, back-up lights, and the new Coronado Deck accessory. In a Ford "first" a limited number of these "Golden Jubilee" replicas were offered to the public through Ford dealers, as were thousands of toy replicas.

1953 FORD PACE CAR

William Clay Ford, 28, test drives the specially prepared 1953 Ford Crestline *Sunliner* Convertible which he drove to pace that year's Indianapolis "500" race. It was the first time a Ford was selected for such an honor since 1935.

A Ford-built passenger car has been selected as the pace car to lead the field across the starting line at the 37th annual 500-mile Indianapolis race. Driver of the 1953 Ford *Sunliner* Convertible pace car will be William Clay Ford, Company Director and grandson of the late Henry Ford. Wilbur Shaw, (three-time winner, and) President and General Manager of the Speedway, will ride with Ford on the pace lap.

The *Sunliner* will be presented to the winning driver at the annual dinner following the race.

Color of the Ford *Sunliner* is *"Pace Car White"*, a new creamy white exterior paint. The car will have special trim features, centering around Ford's *"Anniversary Gold"* as an interior color, marking the observance of the Company's 50th anniversary this year.
FORD ROUGE NEWS, *January 30, 1953*

Lucille Ball and Desi Arnaz made this *Yosemite Yellow* '53 Mercury Monterey *Convertible* famous in their 1954 film "*The Long, Long Trailer*". The car stars in the hilarious comedy when the couple takes a cross-country vacation towing a house trailer and have one dumb adventure after another, from getting the trailer stuck on a Rocky Mountain "shortcut" to trying to park it in a relative's front yard. Through it all, the Merc' never failed.

Behind the wheel of a sporty '53 Mercury Monterey *Convertible*, a pretty driver poses in front of the Lincoln-Mercury exhibit building at the 1953 Michigan State Fair. That year Ford designated one of these eye-catching Monterey soft tops its milestone 40-millionth vehicle. It came off the line September 9, 1953, creating a lot of publicity for the company's 50th anniversary celebration.

Right: A salesman's dream. She's already sold on this dazzling '53 Mercury Monterey *Hardtop* and all that's left is to get hubby to bring in his trade-in and sign the deal. The two-tone beauty listed for $2,300 which included curb feelers and fender skirts, standard equipment on the Monterey.

Above: Ford's new '53 "big-front" *F-100* marked the beginning of one of the most popular American pickup series of all time. Pictured at the 1953 Chicago Auto Show is a *Deluxe* model with chrome grille trim, optional two-tone paint, whitewall tires, and other accessories. To make it more streetable, for the first time, you could order your Ford *Pickup* with an automatic transmission.

Right: Ford's ideal upscale suburban picture of the '50s; the Joneses second Ford, a new '53 *Country Squire* parked in the drive. For these folks, the handsome station wagon with its sporty side panels, and passenger car look and ride was the ultimate status symbol.

Top right: Ford's biggest-selling station wagon in 1953 was the low-priced *Ranch Wagon*.

"We find more people building summer cottages and having to get to and from those cottages with three or four children and a load of baggage, bicycles, groceries and the canary cage. We find the bulk of our (wagon) sales going to the one-car family." HENRY FORD II, 1953

81

"... the flavor and the grandeur of the 2,000 mile dash across Mexico... The long straights, with the cars roaring across the plains under a burning sun, the tortuous climbs through the mountains, the crowds lining the roads near every village, even the buzzards circling overhead; they were all part of it..." LEO LEVINE, FORD, THE DUST AND THE GLORY

Ford Motor Company's powerful Lincoln V-8s would dominate the big car division of the unforgiving "Carrera Pan Americana" Mexican Road Race for three straight years 1952-54. In 1953 they swept the first four places, beating out two Chryslers and a Packard. Three other Lincolns all driven by Argentinians finished in the top ten. The course extended the length of Mexico, over mostly rough terrain from sea level to 10,000 feet. Some of the Lincolns competing in the 1953 race, including Walt Faulkner and Chuck Daigh's third place car No. 53, are pictured with their competitors at one of the overnight stops.

To celebrate its *Fiftieth Anniversary* Ford Motor Company produced nearly a dozen show cars in 1953, including the fabulous Lincoln Capri *"Maharaja"* Convertible which made its first appearance at the Chicago Auto Show. Finished in *Pearl White*, with white metallic leather seats and plush mink carpets, the stunning car was lavishly trimmed in 14K gold. The Lincolns this year boasted a 205 hp V-8 and optional power brakes, power steering and a first in the industry — 4-way power seats. For a heavy car it could do zero to 60 in 12 seconds.

1954 *A MILESTONE V-8*

SOMETIME in the early 1950s a survey showed that the popular image of a Ford with its exciting V-8, was fast, masculine and blue collar, while a Chevrolet, with its decidedly boring 6-cylinder engine, projected a more conservative, middle-class image.

That comparison was about to change forever as Chevrolet announced it would introduce a powerful new overhead valve V-8 for 1955.

There were few secrets in Detroit, so the news came as no surprise to Earle MacPherson, Ford's eminent Chief Engineer. He had introduced that same type V-8 on his Lincolns in 1952, along with ball-joint suspension, and was planning to introduce both mechanical improvements on his Fords for '54. The Chevy V-8 announcement just meant the race was on! The fact was, Ford had already beat it off the starting line with a full year's lead.

But it did come as a shock to the multitudes of old "flathead" Ford V-8 fans that their beloved engine, the one that had been the heart and soul of Ford cars since 1932, had come to its end.

"Plenty of dirty-hands car enthusiasts will shed a tear for the wonderful old V-8," bemoaned *Popular Science* magazine . . . "But if it had to go, as clearly it did, it couldn't have been replaced by a sweeter engine."

Maybe so, but you couldn't tell that to the old Ford mechanics who couldn't get at the plugs on the new '54s without burning some skin.

The new Ford overhead valve V-8 was controversial but it was definitely the engine of the future. Called the "Y-block", because of the shape of its deep block construction, its short-stroke design could deliver smoother performance and longer life than the old flathead. More important, it could develop the extra power needed for some of the new Ford options, like power brakes and power steering.

The new V-8, and ball-joint suspension, made 1954 a milestone year for Ford engineering. But, otherwise, the look of the car hadn't changed much. "You'd have to be an owner or a small boy to tell it from the '53, just by looks," wrote Frank Rowsome, Jr., describing it for *Popular Science*.

But there were a few new styling details worth noting. The body side moldings, deck lid latch, and instrument panel were all new. And the taillights were bigger.

The '54 Fords were also coming into the showrooms in newer, prettier colors, and loaded with extras like whitewalls, radios, and tinted glass. Buyers were tending to want fancier styling, more convenience and comfort. Dealers saw this and were ordering more cars with automatic transmissions, power seats and windows, and such expensive add-ons as continental kits and special-order chrome spoke wheels. Customers were buying more convertibles, and the exciting new Ford Skyliners and Mercury Sun Valleys.

Remarkably, with all this business and the new V-8, Ford still couldn't catch Chevrolet in the sales race. After running in a dead heat all year long, Chevy pulled ahead to win by 17,453 cars — 1,417,453 to Ford's 1,400,000, amid accusations that Chevrolet had unfairly swelled its figures by getting its dealers to register inventory that had not, in fact, been sold.

This raised the possibility that the trusty Ford may have been winning in the sales game all along. ◆

Left: **A showgirl with the all-new glass-top "Skyliner", one of the Ford highlights of 1954, which included all-new ball-joint suspension and the first completely new V-8 engine in 22 years.**

"We figure that 80 percent or better of all car purchases are decided on by women. A woman is naturally style-conscious from birth – in her home, her clothes, everything she does. So when she and her husband go out to buy a car, she wants beauty on wheels."
GEORGE WALKER, FORD STYLIST

At Jacksonville, Florida a pair of southern beauties try out *"America's favorite convertible"*, a gorgeous '54 Ford *Sunliner* in *Sandstone White*. Sun lovers who wanted a little more style in their '54 Ford convertible now had a choice of four top colors and on black tops an optional green-tinted plastic insert over the front seat with a snap-on sunshade.

1954 FORDS DEBUT

... the smooth silky "GO" of Ford power!

A single strip of chrome running the length of the car, as seen on this pretty *Sunliner* convertible, gave the new '54 Fords a longer, speedier look. But the big change this year was to the chassis, with an all-new overhead-valve V-8 engine, ball-joint suspension, heavier brakes, and a stronger twist-resistant frame with five cross members.

"Ford was the first to offer a V-8 engine in the low-price field, and although the majority of expensive cars joined the V-8 parade, Ford today is still the only car in its price class to have V-8 power." FORD DIVISION, 1954

The 1954 Ford passenger cars to be introduced to the public in Ford dealer's showrooms all over the nation Wednesday, January 6, will feature the following Ford exclusives in the high volume field:
- An entirely new 130 hp Y-block V-8 engine with overhead valves.
- A new 115 hp I-block *Mileage Maker Six* engine, noted for economy.
- New ball-joint front suspension.
- The *Skyliner* — an entirely new hardtop model with tinted, transparent plastic roof over the driving compartment.
- A smart new Ford *Sunliner* Convertible with an unusual transparent panel inserted in the top over the front seat to provide overhead visibility.
- Four new power-operated driving assists: Power brakes, power steering, 4-way power seat, power-lift windows.
- Fordamatic transmission.

In its 1954 passenger cars, Ford again is offering three lines, Mainline, Customline and Crestline — with 14 body styles. Two new body types added this year are: the Crestline *Skyliner* with the transparent roof and a Crestline *Fordor Sedan*, featuring luxurious trim formerly found only in costly limousines.

The 1954 Ford passenger cars are readily distinguishable by their new grille with its characteristic Ford center spinner, recessed parking lights and jet-type air scoop. There is a new Delta-wing type hood ornament, an increase of half-an-inch in the height of the crown of the front fenders, and a new diagonal slant to the headlamps. Combined with the single strip of chrome running the length of the car, these give the new Fords a longer, speedier look. FORD DIVISION NEWS BUREAU, DEARBORN, JANUARY 4, 1954

NEW OVERHEAD VALVE V-8

The 1954 Ford passenger cars which employees have been assembling since mid-December will be introduced to the public January 6th. Heading the long list of exclusive features in the new models are the much-discussed Y-block V-8 engine with overhead valves and the new 115 horsepower I-block *Mileage Maker Six* engine.

Other Company exclusives in the 1954 passenger car line include ball-joint front suspension on all models and four new power-operated driving assists, power brakes, power steering, 4-way power seats and power lift windows – all as options.

The Y-block V-8 engine has a deep block construction and is of short-stroke, low-friction, high-compression design. It has a compression ratio of 7.2 and 130 brake horsepower, up 20 horsepower over the 1953 L-head V-8. FORD ROUGE NEWS, JAN. 1, 1954

Ford owned the low-price V-8 market, building more than 16-million of its famous "flatheads" since 1932. But other makers were getting into the market so Ford had to upgrade fast with the '54 models, beating Chevy by one year in bringing out a new "Overhead-Valve" V-8 engine. One of the first of these is pictured being lowered into a chassis at Dearborn Assembly, January 4, 1954.

Left: Skilled workers at the Dearborn Assembly Plant use a jig with special calibrations to build the complex new '54 Ford ball-joint suspension front-end. First introduced on the '52 Lincolns, ball-joints replaced the old kingpin design. Ford engineers claimed it gave better cornering and, combined with a new frame, shocks and other improvements, was *"the greatest riding advance since independent front wheel suspension."*

POLICE FORDS

If you received a speeding ticket this year the chances are better than 2 to 1 that the police officer who pulled you over was driving a Ford.

The Company has been supplying special police vehicles for the nation's law enforcement agencies for the last 22 years. The Company's newest safety enforcement vehicle, the *Interceptor* police car, has improved acceleration nearly 20 percent over the 1953 model, and is capable of speeds in excess of 100 miles per hour. It is available only to law enforcement agencies.

The *Interceptor* is powered with a special 160 hp overhead valve Y-block V-8 engine fitted into the 1954 chassis and is available in Fordomatic Drive, overdrive, or conventional transmission. Power features on the *Interceptor* engine include high turbulence combustion chambers, 4-barrel carburetor, a new intake manifold, high lift overhead valves, and a high compression ratio. FORD ROUGE NEWS, JUNE 15, 1954

"After World War II, you got rid of fenders, per se, you got rid of running boards, bumpers became more integrated. But then there was nothing really dramatic. They were all refinements. Then we got into kind of a glitzy (period) . . . with the robin's egg colors."
WILLIAM CLAY FORD

Fresh from assembly, six colorful '54 *Sunliner* Ford convertibles bound for lucky buyers somewhere are readied for shipment at the Dearborn Assembly Plant. The sporty new soft-tops came in 13 shades ranging from *Sandstone White* and *Sea Haze Green* to *Cameo Coral* and *Torch Red*.

Right: The new '54 Ford Crestline Victoria for *"those who want something special. Side windows roll down leaving no center posts."* All Crestline models had a new two-tone "Astra-Dial" instrument panel and a choice of vinyl or nylon upholstery. Customers could now chose from an impressive 156 exterior and interior color combinations, including 13 two-tone combinations.

"Interior beauty once found only in the most costly limousines and bright exterior body colors for new owners who like to express their individuality, are highlights of the new 1954 Ford passenger cars." FORD AD

The novel new *Skyliner* hardtop with the tinted "see-through" plexiglass roof helped boost Ford's image as a style leader. Calling it *"the top hit of the '54 season"*, the company went on to sell more than 13,000 of them. First shown at the Rotunda in Dearborn in January, 1954, two of the snazzy glass-tops were given away to visitors submitting the best "Worth More" features of the new Ford car line. Promoted as the closest thing to a sunny convertible without the wind-in-the-face, the *Skyliner* had a lot of appeal to buyers in snow country but one lady wrote Ford that it was *"...a very weird sensation to be driving through a snowstorm. Makes the car seem colder."*

NEW FORD SKYLINER

The "see-through" theme of Ford's sensational *Skyliner* was extended inside.

"...I (designed) the astro-dial which was a see-through speedometer... to do something on the instrument panel that would perk it up. The astro-dial sat vertically above the panel so you could look through it... and the idea there was to give it a dramatic back lighting." JOHN NAJJAR, FORD STYLIST

Bottom left: Workers install the plexiglass panel on a *Skyliner* body at the Atlanta Assembly Plant.

Ford's completely new 1954 Crestline *Skyliner*, featuring a transparent plastic top over the driving compartment, affords an open-air freedom which gives the illusion of riding in a convertible with the top down but retains the weather protection of a hard-top model.

The Skyliner's side windows roll down and out of sight to complete this wide-open feeling. The strong, molded sheet of plastic is tinted a neutral blue-green to block out sixty percent of the sun's heat rays and seventy-two percent of the glare.

The new overhead visibility of the *Skyliner* permits passengers to enjoy mountain scenery, read passing business signs, or watch for the change of overhead traffic lights through the transparent roof panel. If needed, a sunshade to match the underside of the top lining can be snapped in place.

Covering the entire front seat compartment, the plastic top is mounted on a weather-tight rubber seal and is framed with a bright metal molding. Gold plastic panels at the sides of the rear windows have a waffle design and bear the Ford emblem in colors. The bright metal molding across the top of the rear window is ornamented with black enamel insets and serrated body belt moldings have the words "Skyliner" inset in black enamel.

Exterior single colors for the *Skyliner* are *Raven Black* and *Sandstone White*. Two-tone colors include: *Raven Black* with *Sandstone White* top; *Killarney Green* with *Sandstone White* top; *Sandstone White* with *Cadet Blue* top; *Sandstone White* with *Killarney Green* top; *Sandstone White* with *Cameo Coral* top; *Cameo Coral* with *Sandstone White* top; *Cameo Coral* and *Raven Black* top. FORD NEWS BUREAU, DEARBORN, DECEMBER 12, 1953

Sister model to the Ford *Skyliner* was the new '54 Mercury Monterey *Sun Valley* which was introduced a month earlier. The see-through roof idea came from two Ford experimental cars – the *X-100* and the *XL-500*. Some reactions to the Sun Valley: *"Excellent. A convertible without the wind."*... *"Fine now, but in July and August?"*... *"A lot of glare at high noon."*

Right: A birdseye view of the new '54 *Sun Valley*, first American body-type of its kind, showing the unique transparent plexiglass roof panel tinted green to tone down light, heat and glare. These beauties listed for $2,706 and came with radio and directional signals, and a special choice of colors. There were 9,761 of them built.

The new 1954 Mercurys with their long, low, wrap-around lines and classic hooded taillights were superbly styled, as represented by this glamorous red Monterey *Convertible*. Mercurys this year got new ball-joint suspension and a new 161 hp Overhead-Valve V-8.

Left: The car above, lighted to show its stunning two-tone interior, is photographed for publicity at the Lincoln-Mercury production studios at the new Ford Styling Building in Dearborn.

Right: The '54 Mercurys made their first appearance at 1,800 dealers nationwide on Dec. 10, 1953. A new Monterey *Hardtop* is pictured in one of the 22 two-tone body and interior combinations offered.

The latest equipment to go with a pair of '53 Ford "Big Job" fire trucks at Ladder Company No. 1, in Dearborn, Michigan is a custom-built '52 Ford *Courier* ambulance. The *Courier*, derived from the station wagon body, was a model in the Ford commercial line from 1952 to 1959 and was used mostly for light delivery. Low cost, rugged design, fast response, and reliable performance have traditionally given Ford trucks and commercial cars a big edge over the competition when it comes to winning bids.

Above: The sleek new 1954 Lincoln Capri *Convertible Coupe* at $3,699 came standard with *Leather* interior. Some options were tinted glass, vacuum antenna, road lamps, and power brakes, steering, seats, and windows.

Right: A Los Angeles Lincoln dealer points out optional Kelsey-Hayes chrome wire wheels installed on a '54 Capri *Convertible Coupe*. Kelsey-Hayes was a longtime Ford and industry wheel supplier that fell on hard times after World War II until sports car styling in the 50's stirred a cosmetic interest in wire wheels. Seeing an opportunity the Detroit company got a new surge of business by supplying a limited-production 40-spoke wheel as an expensive accessory for Ford and other makes, primarily Cadillacs, Buicks, Packards and Chryslers.

Left: Ray Crawford's winning Lincoln Capri in the large stock car division of the 1954 Mexican Road Race makes a pit stop for fuel and brakes. At times reaching 130 mph, it streaked across the finish line at Juarez, near the U.S. border at El Paso, Texas, to complete the five-day 1,912-mile course in 20 hours and 40 minutes. Crawford, a Pasadena, California grocery magnate, raced the Bill Stroppe-prepared car as a private entry, beating seven other Stroppe-tuned Lincolns sponsored by Ford Motor Company. All the Lincoln entries, according to Ford were *"strictly stock in every detail, with the exception of safety modifications"*, which, in race lingo, meant a "tweaked" engine, shot-peened steering arms and spindles, induction-hardened rear axles, and heavy-duty sway bars and tires.

Above: A Chevy leads a '54 Ford entry from Argentina on a stretch of the 1954 Mexican Road Race. This was a controlled section. Most of the race ran over dirt roads and through villages at high speed where the cars sometimes hit pigs, chickens, and people – leading to the races being banned in 1955 because of all the fatalities to drivers and spectators.

Canadian 1954 Ford cars were different than the U.S. models, with special trim. The Canadian Mercury was called the *Monarch*, and the Monterey the *Lucerne*. Models pictured at the 1954 Montreal Auto Show include the new Lucerne *Sun Valley* and *Convertible*, Ford *Sunliner* Convertible, and *Victoria*. Canada began making Fords in 1904 and opened its biggest operations at Oakville, near Toronto, in 1953. Its original old factory opposite Detroit became the famous Windsor Engine Plant, building V-8s for both the U.S. and Canada.

At the near right is Ford Dearborn's *X-100* experimental car, newly arrived from the 1954 Paris Auto Show on its world tour. Some of its advanced ideas were the sliding transparent roof canopy, aluminum and magnesium alloy body and engine, 12-venturi carburetor system for variable driving speeds, four over-lapping windshield wipers, electrically operated jacks for each wheel, and such conveniences as dictaphone, ship-to-shore telephone, and electric shaver.

1955 FORD GOES SPORTY

AMERICANS were becoming fascinated with fast imported European sports cars so it was not by chance that the 1955 Fords had some of their sculpted speed lines and low-slung, fun-loving looks. After all, this was the year Ford brought out the companion sporty Thunderbird.

"We were just beginning to get into an era," recalls Bill Boyer, who was at the time of the '55 car's development, working on both the design of the secretive new Thunderbird and the Mystere experimental car, "where people were replacing the wornout wartime cars and were beginning to look for a little pizazz and some excitement . . .

. . . We were trying to get Ford out of the Tin Lizzie era; trying to appeal to the people that wanted to go upscale and yet wanted to be youthful. Ford had the reputation of being the blue collar worker's car. Mercury was the blue collar worker's Buick, and it was that kind of imagery we were trying to overcome . . ."

Frank Hershey, who went to work for Ford Styling in 1952 and was later Ford Studio Chief Stylist, worked on the '55 Ford design, which amazingly while it had a new chassis, wasn't a new body style at all but a clever sheetmetal rework of the old '52-54 model.

"We made that wild molding to get us by one year," he says in his memoirs, referring to the car's most famous feature, the Fairlane's V-shaped "checkmark" or "swash" body side strip which sweeps along the side.

Hershey thought the "swash" borrowed from Boyer's Mystere was just buying time until they designed the '56 model, but to his surprise it was the one styling feature people liked most so they kept it on for another year.

Like all the other auto makers in 1955, Ford went after the fast-growing youth market, making it easier to buy a new car and promoting performance. By now, speed had become a national mania. Auto fans were flocking to sports car and stock car racing events. Drag racing that used to take place on backstreets with a lookout for cops had become legitimized into a weekend rite of organized contests on old abandoned roads and landing strips. Speed magazines such as *Hot Rod, Rod and Custom, Road and Track,* and *Motor Trend* helped fuel the craze.

To compete with Chevy's much publicized new V-8 engine, Ford offered for the first time, a $35 special "Engine Power Package" with a four-barrel carburetor and dual exhausts that pushed the regular 272 cubic-inch V-8 to 182 horsepower. Later, to keep speed fans from straying, came hotter 198 and 205 horsepower versions of the 292 V-8 that in the '55 Ford could be taken straight to the drag strip, raced, and cruised back to town in time for a movie – all without lifting a wrench.

Other mechanical "firsts" for Ford this year were tubeless tires, gasketless spark plugs and air conditioning for V-8s.

The Ford line-up was full of new ideas in 1955. Besides the Thunderbird "personal sports car", this was also the first year for the 4-door hardtop for families who wanted a little more style, and the first year for the *Crown Victoria* for people who wanted the ultimate hardtop.

Presented in a stunning array of colors, the new '55 Ford models were unveiled in 6,500 dealer showrooms across the nation the week of November 15, 1954. The timing and all the press comparing the Ford and Chevy V-8s couldn't have been better. The appetite of the American people for new cars was already turning out to be bigger than was ever thought possible. Population was growing, consumer income was rising, and more families than ever were in the market for a new car.

The man who really shaped the '55 Ford, at least pulled all the right ideas together, was once again – George Walker. In May, Henry Ford was so impressed with the outside consultant's uncanny styling instincts that he hired him to work full time for the company, along with his gifted assistants Joe Oros and Elwood Engle. Walker was immediately named Vice-President and Chief of Styling. Oros became chief stylist of the Ford studio in 1956, and Engle became a top stylist for Lincoln-Mercury.

The 1955 Fords were so right and so well received that by the end of the model year more cars were sold than in any year in the company's history. In fact it was the best year ever in the American auto industry. And, while it was close, Ford still took second to Chevrolet in the sales race. ◆

Right: A '55 Fairlane *Sunliner* at poolside. The sporty-looking 1955 Fords were perfectly timed. With a booming economy and lots of leisure time, Americans were looking for more fun in their cars.

THE ALL-NEW 1955 FORDS

Ford's new Fairlane series proudly leads the four lines of 1955 Ford passenger cars, all featuring modern styling inspired by the Thunderbird, wrap-around windshields, a lower silhouette and more powerful engines.

The Crown *Victoria*, one of the five models available in the Fairlane series, is the first Ford sedan under five feet in overall height. Featuring an arch of chrome over the top, it is available with either an all-steel top or a transparent plastic roof over the front seat.

In addition to the two *Crown Victoria* styles, the Fairlane series includes the *Sunliner* convertible, the *Victoria*, the *four-door Town Sedan* and the *two-door Club Sedan*.

All Fairlane models are immediately recognized by their distinctive chrome trimlines. They start at the top of each front fender at the headlight, follow the curvature of the fender downward and then sweep back along the side to the tail lights.

Other highlights of the new models include:
- Dual exhausts, which eliminate engine back-up pressure and provide extra power, are standard equipment on all Fairlane V-8 and station wagon models.
- Tubeless tires, standard equipment on the new Fords, are safer because they retain air longer after a blowout or puncture
- Ford's new special high-compression V-8 engine, available with Fordomatic drive on Fairlane and Station Wagon models, is rated at 182 hp.

For 1955, Ford has made a separate series of its popular station wagons and has expanded the line to include five models rather than four, all with steel bodies. Models include the eight-passenger *Country Squire* with side moldings of wood-grained glass fibre, an *eight-passenger Country Sedan*, a *six-passenger Country Sedan*, a *Custom Ranch Wagon* and a *Ranch Wagon*.

This year Ford's Customline series includes Fordor and Tudor Sedans. The chrome molding along the sides of the Customline models provides clean, classic lines. FORD MIDWEST PUBLIC RELATIONS OFFICE, CHICAGO, NOVEMBER 29, 1954

Inspired by the low-slung Thunderbird, the all-new 1955 Fairlane, represented here by the *Victoria* model, brought a sophisticated "going places" look to Ford cars. Instantly recognizable by the daring swash or "checkmark" side molding, the new Fairlanes showed a growing European sports car influence in American automotive design, such as visored headlights, and latticework grille. Styling touches that would quickly become Ford hallmarks.

A game of tennis is the perfect setting to show how much fun you could have owning a sporty new '55 *Sunliner* convertible finished in dazzling *Goldenrod Yellow*. Ford surveys showed that convertible buyers had a higher mix of women and single people and tended to be younger and better educated, with higher incomes than the average car buyer.

Only Ford delivers "Trigger-Torque" power and Thunderbird styling!

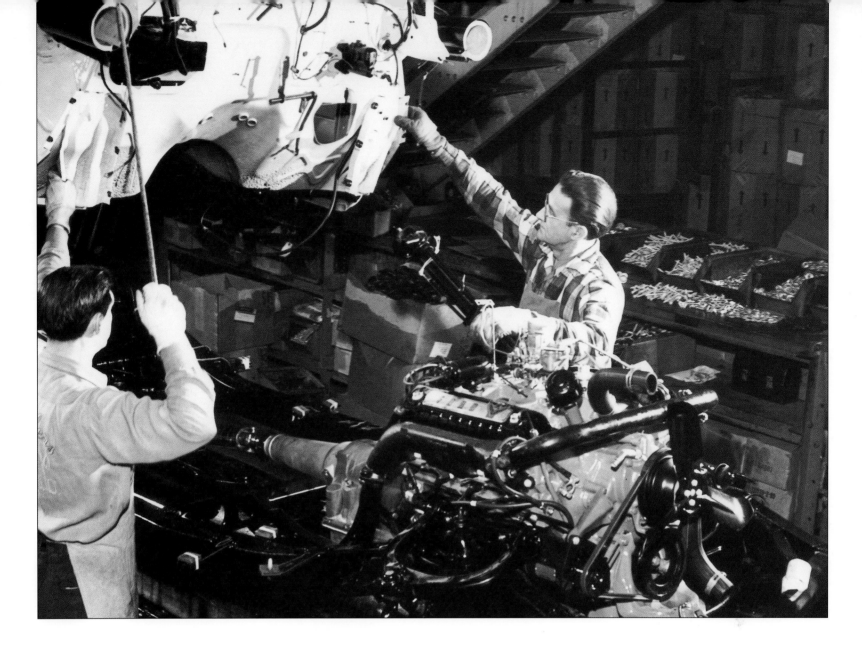

Mating a body to a V-8 chassis at Dearborn, January 5, 1955. Ford's new teletype system made possible the precise arrival at assembly points of all the new combinations of engines, colors, and options.

Left: In an operation little different in method from Model T days, brawny assemblers at Dearborn "drop" a '55 Ford V-8 engine from an overhead trolley into a sprayed black chassis moving down the assembly line. Times had changed considerably. In old Henry's day, anyone caught smoking on Ford property got the boot.

ORDERS BY TELETYPE

Because of a new system of sending messages, the Dearborn Assembly Plant is insuring that dealer orders for cars are filled more quickly and accurately.

The system is comprised of teletype machines. They have partially replaced TelAutograph, an electrically controlled relay system which formerly transmitted production scheduling information from department to department.

By using the teletype, arrival of the right part in the right place at exactly the right time is assured. As a result, cars rolling off the line conform with dealer's requirements as to body style, color, special equipment and a multitude of other items.

There is one master sending machine for the final assembly line. It transmits the necessary information to 12 different departments, such as frame, rear axle and motor dress-up. FORD ROUGE NEWS, JUNE 24, 1955

A loaded "special order" 8-Passenger *Country Sedan*, pictured at a Los Angeles auto show in 1955. Fender mirrors and heavy-duty roof rack suggest there may be more surprises to this classy wagon. Just the hauler for a Western movie producer's use out on the range, a stage for the mountain lodge, or maybe a hunting car for the well-heeled sportsman.

Left: Baked on finish barely cooled from the ovens, a Fairlane *Sunliner* convertible body leads other models on the trim line at Dearborn Assembly in early 1955. Markings in grease pencil give assemblers the build details. The bodies were built-up in welded sections from steel stampings that came from here at Dearborn, as well as from Ford plants at Cleveland, Ohio and Buffalo, New York.

SPECIAL ORDER CARS

Every once in a while in the course of his duties, Don Giese, Driveaway Garage driver, takes an odd-colored passenger car or Courier off the final line at the Dearborn Assembly Plant.

The color tells Giese that the car is a "DSO" — A *Domestic Sales Order* unit built to the specifications of a private business firm or perhaps a government agency.

Most common of the DSO colors are the bright reds and yellows of breweries or bottling firms and conservative blues and greens of utility companies and the armed forces. Bodies of the units are painted in the plant's Paint Shop. Finishing work, such as affixing decals, stencilling, or stripe painting, is applied in the Driveaway Garage.

While it is the most obvious feature of DSO's, color alone does not always make the unit a special one. A black sedan, for example, might boast the powerful Interceptor engine and heavy equipment of a car built for a police department.

Special equipment frequently must be built or purchased and scheduled into production. Examples of purchased equipment are windowless rear doors and steel mesh screens used on Couriers built for a tobacco company. Except for Sunliner, Victorias and Thunderbirds, every Ford body type is represented in DSO production. One popular order is for an ambulance built from a regular Country Sedan with a Courier side open-rear door. FORD ROUGE NEWS, AUGUST 19, 1955

'55 CHICAGO AUTO SHOW

Seven of the 16 Ford body styles available for 1955, plus the famous *Thunderbird* personal car, will be on display at the Ford Division exhibit space during the Chicago Automobile Show.

Highlighting the Ford exhibit, on a slowly moving turntable, will be the entirely new *Crown Victoria* with its tiara-like arch of chrome over the top. Exterior colors of *Tropical Rose* and *Snowshoe White* are complemented by the interior trim of magenta and white vinyl. The chrome tiara motif over the roof is carried through on the inside.

The Fairlane *Victoria* to be shown features exterior colors of *Regency Purple* and *Snowshoe White*, with interior trim of magenta and white vinyl. The Fairlane *Sunliner* convertible, in *Goldenrod Yellow* and *Raven Black*, features an interior trim of black with a yellow vinyl insert.

A CinemaScope movie of Ford cars and trucks will be featured on the main floor just to the left of the Halsted Street entrance to the Amphitheatre. Just beyond the CinemaScope theatre will be a special "Thunderbird Salon" featuring three of the stylish, high-powered personal cars.

Cutaway models of the three Ford passenger car engines for '55 – the I-Block Six and two Y-Block V-8s with respective horsepowers of 120, 162 and 182 – will be on display along with models of the various optional power assists. The 182 horsepower engine is available only with the new Fairlane and Station Wagon series.

Narrators will be on hand during all show hours to tell the story of Ford's colors for 1955 at the revolving "Fashions in Fabrics" exhibit, new this year. MIDWEST PUBLIC RELATIONS, FORD MOTOR COMPANY, CHICAGO, NOVEMBER, 1954

Models narrating the Ford exhibit at the 1955 Chicago Auto Show pose with a glamorous new Fairlane *Sunliner* Convertible in *Goldenrod Yellow* and *Raven Black*, with sporty accessory Sports Wheel Carrier and wire wheel covers. On the turntable, beyond, is the featured Fairlane *Crown Victoria* in *Tropical Rose* and *Snowshoe White*. Purred the girls to their admiring crowds, "... Trim, long fender lines and sophisticated 'going places' flare gives Ford that years-ahead look ... buy a Ford and ride in style ..."

Right: Another view of the Ford exhibit at the 1955 Chicago Auto Show includes a Fairlane *Victoria* in the foreground.

The beautiful new '55 Fairlane had, according to Ford, *"all the desirable features we know how to put into a car... Inside, you sit in the lap of luxury, surrounded by color-and-fabric combinations of distinctive taste and quality..."* The Fairlane was certainly a bird-of-paradise contrast to anything in the past. Optional air conditioning on Fords for the first time, and a Fordomatic which shifted down for faster getaway by pressing the gas pedal to the floor were some of the new selling points.

Left: Accounting for nearly one out of three Ford Victoria sales, the distinctive '55 Fairlane *Crown Victoria*, with the tiara-like arch, also took sales from Buick, Oldsmobile, DeSoto, and other cars in the medium price bracket, which was the design objective. Of the 33,165 built, 1,999 customers ordered their "Crowns" with the optional "Glass Top" roof insert over the driving compartment.

Above: An inspector at Dearborn Assembly checks off a trio of wild-colored '55 Ford Fairlane Crown Victorias. Two-inches lower than the other Fords, with a stylistic "basket handle" band of chrome running across the top, *Time* called it Ford's "gaudiest car", but today it is one of the hottest collector items. It cost about $100 more than the regular *Victoria*.

"...The new Ford paint combinations are dazzling, e.g., a white and lavender hardtop with orchid interior."
TIME MAGAZINE

"At the 1953 Paris Auto Show, Henry Ford II was admiring the low-slung Jaguars, Mercedes, and Ferraris, when he turned to George Walker and asked: 'Why can't we have a sports car like that?' Walker was waiting for just such a chance, his staff had been working on a sports car for months. He made a quick transatlantic phone call, and when he and Ford got back to Detroit a clay model of the Thunderbird was waiting. "Instead of a sag like a Jag," says Walker, "it had a clean, straight-line treatment that was typical of other Fords at that time. We wanted to get a small, sporty car without making it look small, since the American likes a good-sized package for the good chunk of money he pays." TIME MAGAZINE

". . . We had two lines, one for the passenger cars and one for the new Thunderbird and as I remember the number of cars per hour was nine versus the passenger car (66) . . . and the company took great pains to assure that the quality supported the price . . ." SOUREN KEOLEIAN, DEARBORN ASSEMBLY STAFF

". . . We started on the body and got it well along and then ran into the usual problems . . . we can't afford all the tooling . . . so . . . let's plagiarize the '55 Ford with its finials and its headlamps and taillamps . . . We finished up the 'Bird in '53. We were completely finished with the 'Bird probably two years prior to production." BILL BOYER, FORD STYLIST

NEW THUNDERBIRD

The first public unveiling of the sensational new Ford Thunderbird was at the Detroit Automobile Show in February, 1954. Henry Ford II, the 36-year-old president of Ford Motor Company, couldn't be more pleased to demonstrate it to the press. He had other reasons to be proud. In ten years he had brought the Company through a remarkable transformation.

Left: A 1955 Thunderbird body meets its chassis on the line at Dearborn Assembly, which was the only plant that built the car. Full production of the sporty 2-seaters began here in September, 1954.

Ford's new personal car, the Thunderbird, which was only a "dream car" at major auto shows a year ago, now is in full production and on the highway in all parts of the country.

The Thunderbird, with its all-steel body, is the first "personal car" of American manufacture. Ford engineers designed it to combine high performance with the comfort, convenience, and safety of a conventional car. Although it is just four feet, four and 2/10 inches high in the hardtop model, the Thunderbird will accommodate three people, with rear compartment space for their luggage.

The adjustable steering wheel slides in or out three inches, and can be locked in the position the driver prefers. The power seat is standard equipment.

Power steering, power window lifts, and power brakes are available at extra cost — the same convenience features offered in the full Ford car line. Roll-up windows and the hardtop give the Thunderbird year-round utility in any climate.

The car's high performance is provided by a Ford Y-Block V-8 engine of 292 cubic inches displacement with 3.75 inch bore and 3.30 inch stroke. Compression ratios are 8.1:1 with standard transmission and 8.5:1 with Fordomatic. The engine is rated at 198 horsepower with Fordomatic, and at 193 horsepower with standard or overdrive transmission. Fordomatic and overdrive are optional at extra cost. NEWS BUREAU, FORD DIVISION OF FORD MOTOR COMPANY, DEARBORN, NOVEMBER 29, 1954

Above: Everybody wants a closer look at the new Thunderbird at the 1954 Michigan State Fair.

Left: The beautiful new '55 Thunderbird draws a crowd at the Chicago Auto Show. Ford's answer to the 1953 Corvette, it had a V-8 instead of a Six, a real steel body instead of fiberglass, had classier styling, a choice of hard or soft tops, and was far more appealing to the country club set. It was named for the thunder and lighting mythical bird of the American Indians of the southwest. More than 5,000 names were considered before "Thunderbird" was picked.

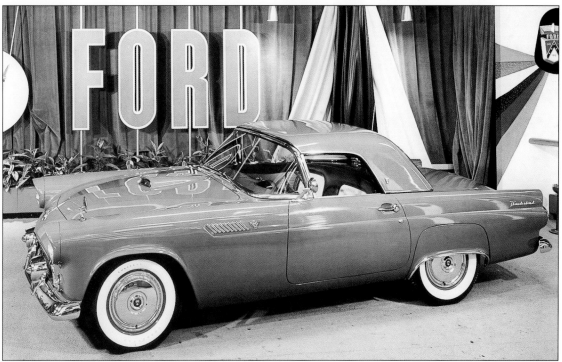

Outside the Ford Factory Delivery showroom in Detroit, Ed Patterson of Ford Customer Relations shows Mrs. O'Dell how easy it is to open the tailgate of her new '55 Ford *Country Sedan*. Factory buyers were usually on vacation from out of state, arriving at Willow Run Airport or by train.

FACTORY DELIVERY PLAN

In a time when visitors could go to the assembly plant and watch their car being built, Ford encouraged buyers to come to Detroit and drive away their purchase. Here, Mr. and Mrs. Ury O'Dell sign for their new 1955 Ford at the Factory Delivery showroom located at 1833 East Jefferson, near downtown Detroit.

Left: At the Factory Delivery lot next to the showroom above, specially-ordered new '55 Fords await buyers to take them home. The more expensive Fairlane and station wagon models predominate.

From all parts of America, an average of 50 visitors arrive daily in Detroit to get their new Fords through Ford Division's *Factory Delivery Department.*

Most of the visitors combine vacation, pleasure or business with the trip to accept direct factory delivery cars they have ordered from their home-town Ford dealers.

The special department is located less than a mile from Detroit's downtown hotels and office buildings. John F. McGuire, factory delivery department manager, explains the service with this example:

"A resident of another city orders the exact model of Ford he wants from his home dealer, who arranges to have us deliver the car. About three weeks are needed to schedule the car and build it to the customer's exact specifications at our Dearborn Assembly Plant. The system is set up so the customer knows the exact day and hour when his car will be ready. He can time his trip to Detroit so he will have the use of his car for his entire visit here, as well as the return trip.

"Usually it takes only an hour to complete the transaction, and the customer drives away in his new Ford."

The Delivery Center is equipped with a lounge where customers wait while necessary papers are cleared. Television, free coffee and soft drinks help guests pass the time. To plan the homeward trip, free state maps and travel advice are available.

Many visiting buyers take free tours of Ford's Rouge Plant, the largest in the world, and the Ford Rotunda in Dearborn. FORD NEWS BUREAU, DEARBORN, NOV. *29, 1954*

"... It's the woman who likes colors. We've spent millions to make the floor covering like the carpet in her living room." GEORGE WALKER, FORD STYLIST

Above: A red-hot line of completely redesigned new Mercurys for 1955 and a family already sold on a gleaming Montclair *Hardtop Coupe* makes this salesman's job easy. The Montclair class, which also included a *Sun Valley* "glass top", and a convertible, was Mercury's new top-of-the line series for 1955, succeeding the Monterey. Two new Mercury options this year were power lubrication and a device that automatically started stalled engines when the Merc-O-Matic selector was moved to neutral.

Top right: The new '55 Mercury Montclair *Hardtop Coupe* was a winner with the sophisticated ladies who could see themselves riding in style. At $2,400, it was a $400 step-up from the sporty Ford *Victoria*.

Right: A 1955 Lincoln Capri *Special Custom Convertible*. The new Lincolns were longer than the '54s, with raked headlights and canted taillights and such mechanical improvements as a new "Turbo-Drive" transmission. Leather interiors this year featured moccasin-stitched horseshoe-shaped bolsters.

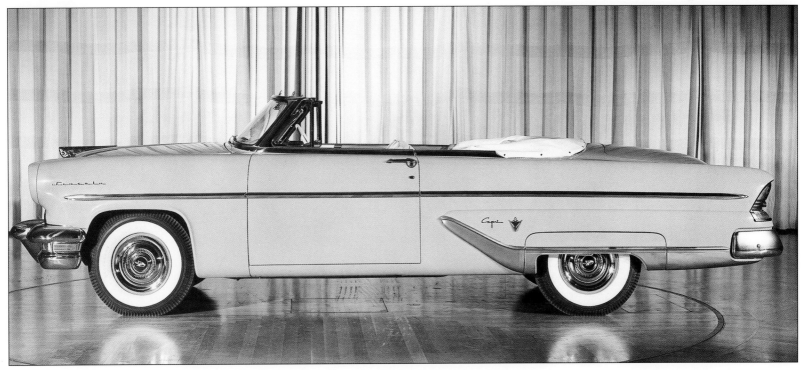

A handsome 1955 Ford *F-100 Custom Cab Pickup* in *Snowshoe White* at the Chicago Auto Show. New features this year besides the deep-V grille bar were the big one-piece curved windshield, and wider doors and rear window.

Left: A look inside the Ford Experimental Engine "build-up" shop at Dearborn in mid-1955 as every man freezes in position for a publicity shot. Here, engines and modifications came off the drawing boards to be built and rigorously tested before going into production some time in the future.

NEW CUSTOM CABS

Ford's new custom truck cabs are available in two-tone color combinations with many of the same attractive colors that are offered on the company's automobile line for 1955.

Snowshoe White on the cab roof and upper back panel may be ordered with any one of seven basic colors: *Vermilion, Raven Black, Aquatone Blue, Banner Blue, Waterfall Blue, Meadow Green,* and *Sea Sprite Green.* The seven single tones are available on all custom and standard cabs.

Cab upholstery is color-matched to the exterior colors. FORD DIVISION, DEARBORN, NOVEMBER 29, 1954

A time exposure captures the action on the floor just before the opening of the 1955 Denver Auto Show. At the center of an exhibit of gleaming Pontiacs, Chrysler products, and sports cars, a sexy new Ford Thunderbird, with top down and hood flipped open, is already the most talked-about car there. In the foreground of the Ford display, ready for high mountain adventure, is an inviting new Fairlane *Sunliner* convertible.

1956 YEAR OF THE CLASSICS

THERE was never a more promising year for Ford Motor Company in the fifties than in 1956. It was one of those rare vintages when every car turned out to be a star — from the beautiful new Fairlanes and Mercury Montereys, with their youth, power and performance, to the Thunderbirds and award-winning new Continental Mark IIs, with their sporty European flair. Together the cars were such a class act it's little wonder an impressed *Motor Trend* magazine was compelled to vote the whole lineup, "Car of the Year".

The timing was right, even with a small hitch in the nation's economy.

Auto mania was at its zenith. It had so reshaped America with its miles of new highways and rows of motels and shopping strips that by now it was said you could theoretically live a reasonably full life in Los Angeles without leaving your car. You could eat your meals or see a movie at drive-ins, do your banking at drive-in banks, shop for groceries, and have your shoes, radio or lawnmower repaired at drive-in stores. You could even attend church services on wheels.

It was a car paradise where annual introductions and shows were important events not to be missed. The speed craze was bigger than ever. On Saturday nights kids in their hopped-up, customized and personalized cars cruised endlessly, laying rubber and hanging out at the drive-ins ready to drag anyone at the drop of a hat. What you drove was who you were and if you somehow lacked an exciting personality — that didn't matter if you had an exciting car.

Out on the NASCAR circuit Ford was back into company-sponsored racing, tearing up the tracks winning 14 Grand National Division races to Chevy's 3, and bringing in new fans like never before.

It was a terrific environment to sell the powerful, beautifully styled '56 Fords and it could have been the company's best year ever but for Robert McNamara, one of Henry Ford II's original management team of "Whiz Kids", recently made head of Ford Division. Unlike his Chevy and Plymouth main competitors, instead of promoting the car's ability to peel out at traffic lights and how good you'd look in one out on a date in the moonlight, the strait-laced office-type who would become the gloomy U.S. Secretary of Defense, chose "Safety" as the car's marketing theme.

Consequently, in an industry where a dash of chrome or the right advertising word could make a difference in thousands of cars sold, McNamara thought people would come pouring into showrooms to see Ford seat belt "safety" demonstrations, deep dish steering wheels, safety door latches, and dramatic films showing dummies thudding into padded dashboards. The campaign was a disaster.

"There were some rather snide commentary about Ford selling safety and Chevy selling cars," says Ford stylist Bill Boyer. "We were being the good guys selling safety and going right down the tube. It was bad news in '56."

Ford sales were going downhill while Chevrolet widened its lead by nearly 300 percent that year — everywhere but in Pennsylvania. Here, a young district sales manager in Philadelphia by the name of Lee Iacocca had come up with a plan that customers could get a new '56 Ford for $56 a month. He sent his salesmen around to parking lots to find likely trade-ins, make an on-the-spot appraisal, and hang a "wujutak" (pronounced "would-ya-take") card and a packet of potato chips. The card read, "The chips are down. We're selling cars for $56 a month."

The gimmick worked like a charm and as Iacocca rung up the sales, a desperate McNamara decided to try it on the whole country — minus the free potato chips. The "'56 for $56" campaign gave a late spark to sales and probably helped move another 75,000 units — and Iacocca's career up the line, as he was promoted to Dearborn to take charge of truck marketing, and eventually to president of Ford Motor Company.

But McNamara's ill-conceived "Safety" campaign, while a noble thing for the local civic club, handed Chevrolet a victory it shouldn't have won so easily. In 1955 Ford officially sold 65,000 fewer cars than Chevy. In 1956 it sold a quarter-million fewer. ◆

Left: **Ford styling under George Walker reached perfection in 1956 when every car in the line was an award winner. Assembled on a Michigan beach with a Continental *Mark II* and Lincoln Premier (extreme left) are three of the sportier models: a Mercury Montclair *4-Door Phaeton*, Thunderbird, and Ford Fairlane *Sunliner* Convertible.**

The 1956 Fords ... With the priceless added safety of "LIFEGUARD DESIGN."

The showroom of Stuart Wilson Ford in Dearborn, shortly after the introduction of the new 1956 Ford models, features a *Fairlane Club Sedan,* a *Customline Fordor Sedan,* and a sporty Thunderbird with the new spare tire mount. A salesman shows a customer how to use the new optional "safety" seat belts that were being heavily promoted this year.

THE NEW 1956 FORDS

With its lowered top the rakish new '56 Ford *Victoria* was one of the smoothest, most popular designs that ever came out of Dearborn. Snapped up by buyers as soon as they hit showrooms, the sexy hardtops with their combination low-price, speed, and beauty were so hard to resist that by the end of the year they had not only outsold the hot-selling '55 model nearly 3 to 2, but were the second best selling model in the entire 18 car Ford line for 1956! If Ford marketers had chosen an exciting "youth, power, and performance" theme this year instead of their dull "Safety" message, this would have been the car to finally beat Chevrolet in the sales race.

Safety features offered for the first time by any automobile company, power equal to the Thunderbird, and lower body silhouettes are available in 1956 Ford cars which 6,800 Ford dealers place on display next Friday (Sept. 23).

A *"Thunderbird Y-8"* engine leads the power selections available for 1956. It is installed on Fairlane and Station Wagon models, and develops 202 horsepower for Fordomatic, or 200 hp for overdrive or standard transmission. Customline and Mainline Fords offer a Y-8 engine developing 176 hp for Fordamatic, or 173 hp for overdrive or conventional drive.

Styling advances in the Ford for '56 include a new grille with oblong parking lights at the outer ends set in frames which wrap around the fender sides. Body side molding is restyled for Fairlane, Station Wagon and Customline models. Restyled tail lamps and deck lid handles, a large recessed hood ornament, and a completely new instrument panel are offered in all models.

The 1956 *Victoria* is 1 1/2 inches lower than the comparable 1955 model. Two door and four door sedans also have new tops reducing total car height almost a full inch. However, headroom was not reduced since the contour change is mostly along the top's center line, and fabric headlinings are installed closer to the steel top.

A 12-volt electrical system is standard on 1956 models, providing 80 per cent faster engine cranking. Optional features in the 1956 Fords include power steering as well as power brakes, seats and windows. Air conditioning, heaters, and tinted safety glass are available. This year Ford offers a signal-seeking radio. *Ford News Bureau, Sept. 10, 1955*

CONVERTIBLE TOPS

Most of the new Fords assembled here are prepared for delivery to dealers soon after they roll off the final assembly line. That is, for enclosed models.

The convertibles, though, are only partially finished when they roll off the final line and through the inspection stations. At that point they are "dressed" in not much more than their gleaming paint jobs. Except for the instrument panel, front seat, and some brightwork in the rear seat section, the convertible interiors are bare of trim. The remainder of the trim and the all-important top come later, when the convertible rolls down a second assembly line all its own.

Another 2 hours are required to complete the top and interior work of the soft-top models. Most of the trim which is installed is prepared at the sides of the convertible line, including rear seat backs. Rear quarter and front door panels also are prepared on the specialty line. In all, 38 additional operations are required to complete the convertibles.

Most important, at least from the future owner's standpoint, are the eight operations performed in installing the top itself. Top builders place the top in the car's body, cut away excess material, and fit the brightwork needed to enhance the beauty of the car and protect its interior from bad weather. *FORD ROUGE NEWS, JULY 8, 1955*

A '56 Ford Fairlane *Sunliner Convertible* gets its top as it moves along a special line. Lighter-bodied cars usually received the dark-colored soft-top and the dark-bodied cars the lighter one. Ford called this *"Reverse Styletone."*

Right: At the end of its travel through Dearborn Assembly, a '56 Ford *Crown Victoria* in *Meadowmist Green* and *Colonial White* is given final inspection and started for the first time. Just 9,209 of these beauties were built, compared to 177,735 of the regular Victoria. Assembled also were 603 of the rare '56 Crown Victorias with the tinted Glass Top.

"... The completed automobile begins to take shape on the chassis line. Starting with frames, workers add springs, rear axle and front wheel assemblies, steering mechanisms and engines to form the chassis on a moving line. A vast number of parts arrive at the assembly line on schedule by conveyor and other mechanical means ..."
FORD NEWS DEPARTMENT, DEARBORN, MICHIGAN

Left: A Ford quality-control inspector checks the color match of freshly baked '56 fenders headed for the main line at the Atlanta Assembly Plant in December, 1955. The new Fords came in a dazzling array of solid and two-tone color combinations.

Right: A young family takes a tour of Dearborn Assembly and watches the progress of a new *Mandarin Orange* and *Colonial White* '56 Fairlane *Sunliner* Convertible as it comes together on the line.

"... Wherever you drive, you'll bask in the envious glances which Ford's Thunderbird styling draws ..."

Showing off the new car to the neighbors is an old American custom and they didn't come any prettier in the '50s than the '56 Fairlane *Sunliner* soft top. A first for Ford convertibles was the new self-locking top. By pressing a button the top moved up and locked itself against the windshield, so the driver didn't have to reach up and lock it by hand.

Right: Ford dealers from the Jacksonville, Florida sales district inspect the new '56 Fairlane *Sunliner* Convertible. Dual exhausts came standard on the Fairlane V-8s. Nifty add-ons included a Sports Wheel Carrier, *"to dress up your Ford like a Thunderbird"*, wire wheel covers, wheel trim rings, grille guard, deluxe rear antenna, fender skirts, and spotlight.

"It was a wild concept with this swash molding and these pods, and this rather sculptural fin and these big blast tubes coming back, and this bubble canopy with a cummerbund and a snorkel for air conditioning." BILL BOYER, MYSTERE STYLIST

Centerpiece of the Ford exhibit at the 1956 Chicago Auto Show was the futuristic new styling research car, the *Mystere*. From it came the design ideas for the 1955-56 Ford wrap-around windows, the V-shaped body-side molding and two-tone paint scheme, the glass tops, and the "roll-bar" roof band feature of the *Crown Victoria*. The *Mystere* would influence Ford and Thunderbird designs for years to come, including tail fins, quad headlights, and glass areas wrapped into the roof.

Right: A section of the Ford exhibit at the 1956 Chicago Auto Show. In the foreground is one of the deluxe new 2-door Parklane ranch wagons and beyond is a *Crown Victoria* and Fairlane *Sunliner*. In 1956 station wagons amounted to an astonishing 19 percent of all the cars produced by Ford Division and were intensely promoted. By now the eight-passenger *Country Sedan* was the hot model.

Above: The new '56 Mercurys had a softer boulevard ride and slight styling refinements to the grille, taillights and chrome trim. A beautiful example is this sophisticated Montclair *Convertible*. Priced at $2,893, it came standard with curb buffers and dual exhaust high-compression engine *Dual Power Kit*. Options included fender skirts, electric antenna, wheel covers, power seats and windows, and top locking device.

Left: In glamorous two-tones, a *Sport Coupe* and *Convertible* in the top-of-the-line '56 Mercury Montclair series are modeled outside the Ford Styling Rotunda in Dearborn. At the extreme right beyond Oakwood Boulevard in the distance can be seen the Independence Hall tower of the famed Henry Ford Museum.

SPECIAL THUNDERBIRDS

New Thunderbird buyers enjoy "personalizing" the Company's personal car that captured the fancy of America's sports-car-minded public.

"We get all kinds of customer requests, from color changes to shipping a Thunderbird directly to Boy's Town of Italy," says Paul Ehrbar, Supervisor, special Distribution Section at the Dearborn Assembly Plant's Driveaway Garage. Ehrbar said that a well-known British-born TV-motion picture actor ordered a gray paint, regularly used on the Continental, applied to his Thunderbird. The actor also requested that a Thunderbird soft-top be applied atop the car's optional hard roof.

"There are many requests for changes in interior coloring," said Ehrbar. He recalled an eastern ice cream manufacturer who ordered an all-brown interior and a steel industry executive who asked for a special red and black vinyl interior not ordinarily stocked.

Solid *White* interiors were ordered by a famous motion picture family and a long-popular ventriloquist. Chosen for exterior coloring were *Coral Mist*, a shade of light pink, and *Regency Purple*, an all-orchid color. Both colors are applied only on special request.

Ehrbar's Distribution Section follows the special order Thunderbirds as Assembly Plant workers make the required changes on the production line and prepare the "colorful" Thunderbirds for shipment. FORD ROUGE NEWS, MAY 25, 1956

A boyfriend with a sexy new *Silver* T-Bird... for the girl who had everything in 1956...

Priced at $3,297 F.O.B. Detroit, the restyled '56 Thunderbird featured a classy spare tire mount sunk into a rear bumper with dual exhausts. Adjustable steering wheel and tachometer came standard and buyers could chose from three V-8 engines including the new 312 cubic-inch *"Thunderbird Special."* Besides special factory paint including two-tone, some options were power seats and windows, wire wheel covers, a hard top with porthole in lieu of the regular hard top, and a tonneau cover.

A sporty new '56 Thunderbird with optional port hole top, on the turntable at the 1956 Michigan State Fair with a Fairlane *Fordor Victoria* and a *Sunliner* Convertible.

Right: A new '56 Thunderbird, on exhibit at the Ford Rotunda in Dearborn.

"I brought Shirley over . . . we were engaged . . . I said, 'I want you to see this car, I'm in love with it.' . . . and I showed her a flame red '56 T-Bird with the tire on the back and the whole works . . . She got behind the wheel and it was a straight-stick . . . and that was the end of that." LEE KOLLINS, FORD ROTUNDA, SPECIAL EVENTS

143

WIN ON SUNDAY, SELL ON MONDAY! When the new Chevy V-8s began winning stock car races in 1955 and getting big publicity, Ford Motor Company hired DePaolo Engineering to get it back into competition. Indy race legend Pete DePaolo's crew is shown at Daytona, Florida working to prepare the first factory shipment of new '56 Fords for the February action on the beach. Ace Ford drivers were Curtis Turner, Joe Weatherly, and Fireball Roberts. Top DePaolo mechanics were Ralph Moody, John Holman, Chuck Daigh, and Don Francisco.

Right: What Pete DePaulo was to Ford racing in the 1950s, ex-dry lakes hot-rodder Bill Stroppe, from Long Beach, California, was to four decades of racing for Lincoln-Mercury. Here the speed genius hero of the Mexican Road Races, right, and ace mechanic Vern Houle look over their engine work on "Thumper", a factory sponsored '56 Merc that ran 152 mph in the Factory Experimental Class during Daytona's 1956 Speed Weeks. The engine was a highly modified 391 cubic-inch Lincoln V-8 with Hilborn fuel injection and Spalding ignition.

145

"... the biggest treat of the day was watching Curtis throw the car sideways on the asphalt long before he ever got to the corner, then bring it on around in a great shower of sand and commotion, straightening everything out and heading north again while most of the spectators were still flinching, waiting for the imminent disaster which never occurred." LEO LEVINE, *FORD: THE DUST AND THE GLORY*

Left: Ford Motor Company was a big player in 1956 stock-car racing, winning many hot contests in the popular spectator sport. Speeding down the blacktop side of Daytona's famous 4 1/2 mile beach course at NASCAR's first ragtop race held in February, 1956, Fireball Roberts, in his Ford-sponsored, DePaolo-prepared, '56 convertible, leads Bob Pronger's car. Roberts finished 2nd to the legendary Curtis Turner and his hard-charging No. 26 Ford convertible.

Above: Chuck Daigh, preeminent car builder and race driver from Bill Stroppe's shop in Long Beach, California moves up to the starting line for another run with the streamlined '56 Thunderbird he built to win first in class in the "Flying Mile" at the 1956 Daytona Speed Weeks. From a standing start in the sand his car, sponsored by Ford Division, turned 85.308 miles an hour at the mile mark for the winning time.

Left: Another '56 *F-100 Pickup* sale at Stewart Wilson Ford in Dearborn, a friendly dealership on Michigan Boulevard that hasn't changed much through the years. Destined to become more than a popular truck for farm or business the '56 Ford *F-100* would become American youth's all-time favorite pickup to hot rod, mostly because of the big front and visored headlights and roof that gave it that "cruisin'" look. You could order one with a "Six" or "V-8", a choice of five transmissions, and in any of 16 single and two-tone color combinations. A specially trimmed *Custom Cab,* pictured here, and a wraparound rear window were optional. Tubeless tires came standard this year.

Above: Ford offered six station wagon models in 1956. The most expensive was this *Country Squire* that was sporty enough to be the family's main car for work or fun. Side moldings were now simulated wood-grain, made of fiberglass, instead of real varnished wood but were easier to maintain. It came with turn indicators, electric clock, and wheel covers for the suggested list price of $2,512. A V-8 engine at $100, and *Red* and *White* vinyl interior trim for $30 were extra. In 1956 Ford and Mercury combined sold 50 percent of all the station wagons bought in the U.S.

". . . Station wagons, once regarded as a luxury around estates, are today enjoying a popularity among middle income families that is responsible for one of the most spectacular developments in the history of the highly competitive U.S. automobile industry."
FORD, JUNE 18, 1956

Few American cars in the fifties turned more heads than the big 1956 showboat Lincolns. With colors Frank Rowesome, Jr. of *Popular Science* magazine called as wild as *"yesterday's sunset"* they were built on a grand scale to compete bumper-to-bumper with Cadillac. The '56 Lincolns came in the Capri series or the Premier, shown here in the glamorous *Convertible* model. Top speed with the big 285 hp 368 V-8 was 121 mph. It was conspicuous consumption in full flame and the buying public loved it! Sales shot up 84% over the 1955 models.

Left: A window shopper's view of Evans Lincoln-Mercury in Dearborn, Michigan on a summer night in 1956. The featured car is a Premier *Convertible*, just like the one above.

CONTINENTAL MARK II

Edsel Ford brought out the first Continental in 1940 and his son William Clay Ford revived the Mark with the sensational new '56 Continental *Mark II*. Ford spared no cost in its custom production. The car was so smooth it was claimed you could balance a coin on the hood with the engine running. The price with air was a lofty $9,516.

"I was terrific. There I was in Florida with my white Continental, and I was wearing a pure-silk, pure-white, embroidered cowboy shirt, and black gabardine trousers. Beside me, in the car was my jet black Great Dane, imported from Europe, named Dana Von Krupp. You just can't do any better than that." GEORGE WALKER, FORD STYLIST

A modern version of one of America's most admired cars returns to the automotive scene on Friday, October 21, when the new *Continental Mark II* will be introduced in dealers' showrooms across the nation.

Styled with a functional and enduring design, the Continental is produced to meet the strictest quality standards in the automotive industry, according to William C. Ford, vice president and group executive of Ford Motor Company and general manager of the Continental Division.

"The Continental," Mr. Ford said, is designed for an exclusive market — a prestige market — consisting of persons with good taste who want an automobile embodying distinction, luxury, dignity and quality."

The beauty of the car's "modern formal" styling, he added, "lies in its proportions and fundamental composition of line and form, derived largely from the best characteristics of the former Lincoln-Continental." Retained in modern form are several of the most distinguishing features of the earlier car, including a long hood, compact passenger compartment, a distinctive rear roof quarter, short rear deck, and a unique spare tire mount. The tire rests inside the luggage compartment beneath a molding stamped into the rear deck lid.

The *Continental Mark II* is a two-door "hardtop" coupe, available in 14 subdued exterior colors and five two-tone interior combinations and is built exclusively in a plant especially designed for low-volume, high-quality production. CONTINENTAL DIVISION, FORD MOTOR COMPANY, DEARBORN, OCTOBER 5, 1955

A prominent architectural feature of the Styling Building is the rotunda showroom. This structure is 120 feet in diameter. The room is unobstructed throughout its entire expanse and features an arched ceiling some 40 feet high at the center. The room is used for displaying styling models for selection and final program approval . . . FORD RESEARCH & ENGINEERING

Well-dressed "customers" at the Ford Styling Rotunda showroom admire some of the glamorous new models in the Ford Motor Company line for 1956. Clockwise from the top is a Continental *Mark II Sport Coupe*, Mercury Montclair *Sport Coupe*, Fairlane *Victoria*, Thunderbird, and Lincoln Premier *4-Door Sedan*. The showing of cars for top company executives was so early before fall introductions that a '55 T-Bird substitutes for the '56 model which was still going through production changes.

"You can't sell a diamond in a matchbox. You've got to give management what management should have and give it to management in the right setting." GEORGE WALKER, FORD STYLIST

152

153

"There are buyers like the Dallas woman who entered a showroom recently, looked over the stock of Continentals, and left, saying: 'I'll be back'. She returned with a toy poodle and showed the dog a Continental with blue upholstery. The dog jumped in, sniffed, settled down — and closed the sale."
NEWSWEEK, September, 1956

Whole teams of experts contribute their talents to creating the attractive interiors — blending fabrics, plastics and metal into luxurious, yet functional effects. Color specialists may develop and offer for consideration several hundred new shades, grouped in color families, with exterior and interior colors keyed together to present a harmonious whole . . . FORD RESEARCH & ENGINEERING

A Styling Rotunda showing of colorful new Ford Motor Company cars for 1956, pictured at a later date than on the preceding pages, includes a classy production '56 Thunderbird at the center of some of the cars that helped the entire line win *Motor Trend* magazine's "Car of the Year" award that year. Clockwise from the top is a plush Continental *Mark II Sport Coupe*, a dreamboat Mercury Montclair *Convertible*, an exciting Ford Fairlane *Skyliner* Convertible, and an exquisite Lincoln Premier *4-Door*.

1957 AMERICA'S FAVORITE

HENRY Ford II would have to wait a full decade to make good on his promise to beat Chevrolet in sales. The victory finally came with the low-slung '57 models. It was close. Chevy said it won the race. And it may have been decided by buyers as to which car had the neatest tail fins.

Says Ford stylist John Najjar . . . "I don't know where it started . . . but automobile designers started to figure that, well, fins and a lot of flying surfaces, and lines that float any place on the car was the thing to do . . ."

Bill Boyer's Mystere (page 138) was a styling platform to try some of these wild ideas. A masterpiece in the world of futuristic dream cars, it would be the inspiration for the overall styling theme of the new '57 Fords.

". . . The Mystere showcar was done in '54", recalls Boyer, "but . . . the use of the car (in the auto shows) was delayed a year because some of the cues that were used on the '57 Ford were the swash moldings and the fins, which related too much to the production car to expose at that early date."

Joe Oros, the top stylist under outside consultant George Walker says this:

". . . I remember (in 1955) Mr. Walker bringing in Bob McNamara (head of Ford Division) to look at some of my thinking proposals for the '57 Ford . . . I was working on a chalk drawing of the '57 Ford car, and it had the wide, double headlights and the very wide front fenders and the quarter panel fin . . . partly a derivative from a specialty model (Boyer's Mystere) that was being developed in the Ford studio area — a specialty show car.

The next day George (Walker) asked me to begin the model of this proposal that I'd shown Bob McNamara the night before.

. . . So, I started the '57 model in the studio. Frank Hershey thought it was just too crazy and gross. He thought it should be stopped. It was exaggerated in its original start-up, and I may have had the fenders just a wee bit too wide, but I was feeling the car in its development, and I was feeling the front-end relationship to the fenders and the quarters and the whole car as an integral piece . . . The car kept developing and improving, and it was finally approved by management."

According to Ford Chief Stylist, Frank Hershey:

"We weren't getting along fast enough on the '57 Ford. We were trying all kinds of things, and it wasn't working, so I told Damon (Woods, stylist), Let's stay tonight. We'll get rid of everybody, and you and I go and rough it out . . . And we roughed that whole '57 Ford out, and the next day they came and perfected it and that was the '57 Ford."

The conflicting stories as to who styled the car, the company designers or Walker and his men, shows how much the two groups were in competition with each other.

"We had a highly charged environment at that time," says Boyer. ". . . So we had a bit of frenzy there for awhile until George Walker came in as vice-president, which put an end to all that. He could then legitimately say, I'm king of the hill, and you guys do what I say."

In the end the '57 Ford was really a team effort. A car that, as Lincoln-Mercury stylist Gene Bordinat liked to say . . . "went out and did great work!"

It proved to be the recipe Ford had been looking for to beat Chevrolet. Long clean lines, low silhouette, sloping hood, modest canted tailfins like the Thunderbird this year, and of course, more power.

The mistakes of 1956 were not to be repeated. This year dealers asked their customers to come in and "Action Test" the new '57 Ford and check out its "road-ability" . . .

Ford had other successes in 1957, such as the sensational *Skyliner* retractable, but Styling Boss George Walker didn't always hit the mark. In trying to boost Mercury sales, his group styled a flashy, chrome-laden car that year that was too "gewgawed" as *Time* magazine put it, and the entire line got a relatively poor reception.

Gene Bordianat specifically remembers the model that had high expectations: ". . . The Turnpike Cruiser . . . was an exciting enough car . . . (then) the market began to falter a bit, as it frequently does. You know, you're on a high, and then all of a sudden you hit a little, modest dip, and we were eating a lot of Turnpike Cruisers."

But that was just a small setback for what turned out to be Ford's biggest , most profitable year yet. ◆

Left: Destined to become the cars that finally beat Chevrolet in the sales race, hotly-styled new '57 Fords stand waiting to leave for outlying dealers aboard '56 Ford haulaway trucks at Dearborn Assembly.

THE NEW 1957 FORDS

For the first time in its 53-year history, Ford Motor Company is producing two sizes of Ford cars. The new Fairlanes and Fairlane 500's are built on a 118-inch wheelbase. Station wagons, Customs and Custom 300's have a 116-inch wheelbase.

Longer, lower and with the highest performance engines ever offered in the low price field, the 1957 models were previewed by 8,700 employees and their families at the Louisville Assembly Plant on September 30th. Public showing in dealer show rooms began October 3.

Ford's new styling starts with wide hooded headlights and a forward slanting grille, and includes streamlined wheel openings. "Hardtop" styling is the trademark of the conventional two-door and four-door sedans in the Fairlane and Fairlane 500 series. The effect is achieved with thin side pillars. Ford's true pillarless "hardtops", the four-door and two-door Victorias, also are offered in these series.

Special side moldings and ornamentation distinguish each of Ford's five series, which are available in 19 two-tone paint combinations or 12 solid colors. Inside, upholstery is color-matched to the body's paint.

For the first time, a high performance V-8 engine is available as an optional power plant on all Ford cars. The engine, called the *Thunderbird Special*, develops 245 horsepower, and is equipped with a four-barrel low silhouette carburetor.

Standard engine for the Fairlane and station wagon series is the 212 hp. Thunderbird V-8. A 190 hp. V-8 is standard for the Custom and Custom 300 series. Both have two-barrel carburetors. In addition, the 144 hp. *Mileage Maker Six* is available on all models. LOUISVILLE FORD NEWS, LOUISVILLE, KENTUCKY, OCTOBER, 1956

The instantly popular '57 Fords were a dream come true for dealers. For the first time in years they were putting customers names on waiting lists. Here, salesmen at Stuart Wilson Ford in Dearborn, Michigan discuss a newly arrived *Sunliner* Convertible, listed at $2,480 with the 292 V-8. The new Ford interiors featured a "safety contoured" dash with recessed knobs and a smaller-diameter deep center steering wheel.

A crowd takes a look at the new '57 Fords at Cort Fox Ford in Los Angeles. New this year was the Thunderbird-type hood, front-hinged for better access to the engine compartment – and to keep it from flying open on the road.

A Woonsocket, Rhode Island customer was showing his wife and mother-in-law a new Ford on the showroom while the salesman stood back. As he was explaining the features he was constantly being interrupted by the older woman. When he came to the windshield washer he stepped on the foot control and the wash squirted past the windshield, landing on the mother-in-law's head. When the salesman wrote up the sale later, the buyer said half-seriously, "that's what clinched the deal." FORD CREST NEWS

"... Here is the car the whole world has long dreamed about ... the world's only Hide-Away hardtop. Touch the magic button and the Hide-Away roof vanishes into the rear deck ... and you're sitting in the dreamiest convertible under the sun ..." FORD MOTOR COMPANY

RETRACTABLE HARDTOP

The first revolutionary new idea in automotive design since the development of the closed car 40 years ago was unveiled today by Ford Division of Ford Motor Company. It is a hardtop model with a fully retractable steel top which, at the touch of a button slides automatically into the car's trunk.

The functional design of the first automatic all-weather car has resulted in a distinctive appearance previously unattained by any hardtop model. The new car combines the advantages of a hardtop with that of a convertible.

Publication today of photographs of a finished model of the Ford Retractable ended nearly five years of secrecy. In that time, Ford's stylists and engineers perfected a design which has been a long-sought goal of the auto industry. The six-passenger two door car is scheduled for production starting in January and it will be sold by Ford dealers. It will be the featured car at the New York Automobile Show opening Dec. 8.

Operation of the Retractable is deceptively simple. When the driver touches the top button, electric locks are released in the trunk lid and it rises up and back out of the way. Then locks are automatically released in the top, which swings up, back and down into the trunk. The trunk lid closes and locks itself. In 40 seconds, the car has been converted from closed hardtop sedan to an open-air vehicle. To change it back again, the button needs only to be moved in the opposite direction. FORD NEWS BUREAU, NOV. 26, 1956

Introduced at Ford dealers April 18, 1957, the new '57 Ford *Skyliner Retractable* was Detroit's first collapsible hardtop. An engineering marvel that took years to develop with 522 special parts different from other Fords, including relay switches, screw-driven locks, and counter-balancing devices, it was the most talked-about new body style of the year. The car was so out-of-this-world with its top in action that one member of the press, seeing the car for the first time at an auto show, was moved to write ... *"Its hard steel top rears up, slides back and disappears beneath a glossy poop deck as a submarine is embraced by an ocean wave. The effect is awesome!..."*

Santa and his helper with one of the just-introduced new Ford *Skyliner* Retractables outside the 1956 New York Auto Show.

"Yes, there it is... the sleekest, most glamorous convertible under the sun or under the moon... ready to take you straight down the road to adventure..."
FORD MOTOR COMPANY

Left: Workers maneuver a built-up Fairlane front-end onto a 4-door body at Dearborn, November 2, 1956.

Top left: 1957 Ford Skyliners on their own build line at Dearborn Assembly. They were also built at Ford plants at Mahwah, New Jersey; Louisville, Kentucky; Kansas City, Missouri; and San Jose, California. The car's distinctive longer rear quarter was necessary to make room for the roof to slide into the rear deck. To make it all fit, the spare tire on these models was under the floor.

Above: A 1957 Fairlane "500" *Sunliner* Convertible takes shape at Dearborn Assembly. Fairlanes came standard with a 292 V-8. An extra $50 would get you the optional *"Thunderbird Special"* 312 V-8 and – for hot foots – you could get your 312 factory-equipped with a supercharger producing 300 hp. Ford Motor Company was the world's biggest V-8 producer, celebrating the 25th anniversary of its V-8 engine in March, 1957, when it built the 23,441,956th unit.

"I bought a black '57 Ford hardtop from the Company, brand new just before I got married. I watched it being built... I was a trainee at the assembly plant and I had the item number... and I made sure it was built during the day and not on the afternoon shift... and not on a Monday or Friday... So I followed that vehicle to the end... I knew everyone in the plant so when the vehicle hit the final line that car was in outstanding shape..." SOUREN KEOLEIAN, DEARBORN ASSEMBLY STAFF

The time is early morning December 6, 1956, at Ford Motor Company's Dearborn Assembly haulaway yard. The big Rouge Plant, largest of its kind in the world, stands in the background against the gray sky as loads of newly built 1957 Fords of every model, in every combination, prepare to move out on 1955 Ford E & L contract transports to distant dealers. This was a scene that would be repeated many times this day at 19 other Ford plants scattered across the nation as the company rushed to fill the biggest passenger car orders of any year in its history. The volume of demand was so huge that by now the company was buying half the finished parts that went into the cars from outside suppliers. From raw ore shipped in from the Great Lakes the Rouge plant itself produced glass, steel, engines, body stampings such as roof panels, doors and fenders, and bar steel for wire, bolts and springs.

"An exciting car with the look of tomorrow etched into every graceful hood, fender, and body line." 1957 FORD AD

Above: One of Ford's featured cars at the 1957 Detroit Auto Show, a flawlessly-styled *Sunliner* makes it easy to see why the '57 Fairlane was so popular. It had everything: beauty, dash, glamour, hill-flattening power – all for an affordable price. This car has the factory-installed Sports Spare Wheel Carrier, one of the Ford dress-up options this year which included turbine wheel covers, rear deck antenna, rocker panel trim, fender skirts, grille guard, and visored spotlight mirror.

Right: New Ford models on display at the 1957 Chicago Auto Show include a Fairlane *Club Victoria*, foreground, a pair of Thunderbirds, and a *Sunliner* convertible with optional Styletone paint and Sports Spare Wheel Carrier.

Top right: Presenting the new '57 Ford. Its low road-hugging look came from extending the frame to the sides of the car so the floor could be lowered an extra three inches inside the frame rails.

The wives try out a new '57 Fairlane "500" *Sunliner* convertible at a Jacksonville, Florida Ford dealer preview.

FORD CONVERTIBLES

Automobile registration figures show that Ford continues to dominate the convertible market and is outselling its closest competitor by better than a three-to-two ratio.

Sales of conventional Ford convertibles during the first five months of 1957 increased more than one-third over the corresponding period last year. Ford sold 33,623 conventional convertibles this year as compared to 24,633 during the first five-month period in 1956.

In May, one out of 12 Ford cars sold was a convertible – either a conventional convertible with a fabric top or the *Skyliner* with a steel top that retracts into the trunk area. More than 5,000 people have purchased the *Skyliner* since it was introduced two months ago. FORD ROUGE NEWS, JUNE 21, 1957

"There in command of an $11.5 million red brick styling center set in an expanse of playing fountains and shimmering pools Style Boss Walker works at the head of a staff of 650 artists, draftsmen, modelers and engineers. Most are young (average age: 31) all have what auto men call, "gasoline in their veins." Says Walker "You just got to love cars." TIME, NOVEMBER 4, 1957

Flamboyant Ford Vice-President of Styling George Walker, left, reviews the first production 1957 Fairlane "500" *Sunliner* at his styling studio patio. He and his group had designed a beauty! It would give Ford its best year for the convertible, with a whopping 77,728 Sunliners sold.

Right: The "old pea-picker" Tennessee Ernie Ford poses with a 1957 *Sunliner* convertible at Dearborn in late 1956, after signing with Ford Motor Company to sponsor his Tennessee Ernie Ford TV show. His own car was a *Flame Red '57 Skyliner* Retractable delivered from the Louisville, Kentucky assembly plant to his home in Bristol, Tennessee.

A winning combination – a pretty driver ready to try her pretty new '57 T-Bird through the clocks at the Daytona Speed Weeks in February, 1957. She was part of the Ford armada of public relations and news people, factory-tuned cars, engineers, mechanics, drivers and pit crews that had assembled on the famous Florida beach to kick off the race season. The main thing was to dominate Chevy on the track and in the news – which they did handily.

When rumors got out that the terrific new '57 T-Bird would be the last of the classy 2-seaters, sales took a jump to make it the best year for the series. The rumor was a fact. The last 2-seater came off the line December 16, 1957. An instant collector's item the '57 sported an all new longer body, rear fenders canted into fins, more power, more luggage space with room for the spare tire, and new 14-inch wheels. At a base price of $3,383 delivered it came standard with a safety padded dash and crash cushion sun visors. The optional hardtop with porthole was offered in a choice of ten shades and could also be had without the porthole.

FORD SUPERCHARGER

A new 300 horsepower supercharged V-8 engine was announced today for all 1957 Ford cars and the Thunderbird.

Ford Division notified Ford dealers that production had begun on the *Thunderbird 312 Supercharged V-8 engine*, which will be sold as an optional extra feature with a suggested list price of $447.50. The letter stated that because of time required to build up production on the supercharged engines, they would be in short supply until May.

The engine replaces a dual four barrel carburetor Thunderbird V-8 rated at 285 hp which Ford formerly had offered as its most powerful engine choice for 1957. With the new centrifugal type supercharger installation, the engine no longer relies on intake manifold vacuum to draw fuel through the carburetor. Instead, the fuel-air mixture is blown into the cylinders. *Ford Division News Bureau, Dearborn, December 29, 1956*

Ford offered speed fans a factory-installed supercharger for passenger cars and T-Birds in 1957 called the "F-Series". A "blown" Fairlane is shown ready for action with a dual 4-barrel set-up at that year's Daytona Speed Weeks.

Driver Chuck Daigh, in checkered shirt, is pictured on the beach at Daytona, Florida with No. 98, one of two factory '57 "Daytona Birds" he had a hand in building for DePaulo Engineering at Bill Stroppe's shop in Long Beach, California. With bodies lightened with aluminum panels and drilled-holes, a super-charged and modified 312 with Hilborn injectors, Hunt magneto, headers, Halibrand quick-change rear-end and other modifications, they clocked 205 in the Flying Mile here at the 1957 Daytona Beach Speed Weeks.

A 1957 Ford *Country Sedan* takes its place at a Detroit fire station as a mascot to the big Seagraves for fast response to emergencies. Ford station wagons filled all kinds of jobs like this, besides hauling the family. The 1957s — re-engineered like the car — were longer, lower, quieter, easier riding and easier handling than ever. New this year was a wrap-around lift gate and tail gate. When you opened the tail gate, the upper lift gate automatically swung open one third — making it a one-hand operation.

NEW FORD RANCHERO

...It's more than a car... it's more than a truck. It's the new Ford Ranchero!...

A pickup truck that looks, rides, and handles like a passenger car will be unveiled by the company November 12th.

Called the Ranchero, the truck resembles the 1957 Ford passenger car from the front bumper to the vehicle's mid-point. From there on it is a pickup truck with greater load space and lower loading height.

The vehicle will be introduced at Quitman, Georgia at a celebration honoring 21-year-old Wesley Patrick, chosen 1956 Star Farmer of America by the Future Farmers of America. Patrick was promised the first Ranchero off the assembly line when production starts in December.

Instead of a pickup box mounted at the rear of the cab as in conventional truck design, the Ranchero is an integral unit with a completely welded body and a double steel floor for added strength and rigidity. Space for luggage and a spare tire is located behind the seat of the cab.

Designed to provide truck owners with the luxury of a passenger car possessing the utility of a pickup, the Ranchero is offered with either standard, overdrive, or Fordomatic transmission.

Two models are available: the *Ranchero* with a 272 cubic-inch V-8 engine with 190 horsepower and the *Custom Ranchero* with a 292 cubic-inch V-8 engine developing 212 horsepower. Each is also available with the 144 horsepower Mileage-Maker Six. FORD NEWS BUREAU, DEARBORN, NOVEMBER 2, 1956

Taking its marketing cue from the half-car half-pickup "Ute" utility vehicle that was popular in Australia from the early thirties, Ford U.S. created an instant market for the "Ranchero", its own sedan-based version, in 1957. They were offered in either a *Standard* or the *Custom* style with gold anodized aluminum side strip and two-tone paint scheme like the Ford passenger car. By the end of the model year 21,695 of the Rancheros were sold. In Australia this style Ford was made famous for taking the pigs to market on Saturday and "Mum" to church on Sunday.

1957 MERCURY PACE CAR

The *Convertible Cruiser* — a special model 1957 Mercury — has been chosen pacemaker (90-mile-an-hour flying start) for the 41st annual 500-mile Indianapolis Motor Speedway race May 30. An exact duplicate of the car will go on sale through Mercury dealerships across the country this spring. Emphasizing the Speedway theme, the *Convertible Cruiser* has front fender ornaments — featuring a Mercury crest and winged checkered flags. Ornaments are illuminated and serve as directional turn signals. The same theme is carried out in a lighted rear deck lid ornament which has the Mercury head centered between crossed checkered flags. FORD ROUGE NEWS, JANUARY 18, 1957

Styled after the *Turnpike Cruiser* in the Montclair line, the special Mercury "Convertible Cruiser" which paced the 1957 Indy "500" is pictured on display at the Ford Rotunda with some of the winning cars. Bill Stroppe prepared the Merc, extensively modifying its engine to run with the racers. Sam Hanks, who drove Mercury stock cars for the Ford-backed Stroppe, won the Indianapolis 500 that year — and the *Cruiser* pace car.

MARK II CABRIOLET

Rarest of all the Continental Mark IIs was this *Pearlescent White*, with red leather, *Cabriolet* built for William Clay Ford's wife. Pictured at the Chicago Auto Show January 6, 1957, it was originally intended to be the prototype for a new convertible offering that year to rev up interest in the company's struggling line of prestige cars. But high production costs and low demand killed the Continental Mark II. After just 444 cars were built for the 1957 model year the last one was produced May 13, 1957, for a total output of 3,000 units. In later years Mrs. Ford's MKII *Cabriolet* turned up in a Ford employee parking lot sale and was purchased by veteran company engineer Paul Wagner.

Customizing of the standard Continental Mark II hardtop coupe to produce a Continental *Cabriolet* Convertible, has been announced by the Company.

The *Cabriolet* Convertible, which made its debut at the Texas State Fair in Dallas, is being fashioned by the Derham Custom Body Company of Rosemont, Pennsylvania.

It is the first true Continental convertible produced since the Lincoln Continental series was suspended in March, 1948.

All the new moldings required as a result of the conversion are hand-made and covered with chromed metal. The body has been widened at the pillars with windows hand made to follow altered body contours.

Side arm rests in the rear seat have been replaced by a portion of the convertible mechanism while the package tray has given way to a pocket panel to house the top when down, and the rest of the actuating mechanism.

The changes in the rear of the passenger compartment necessitated redesigning the rear seat to eliminate the center arm rest and installation of a single-unit seat cushion and back.

Despite the fact that the convertible top of the *Cabriolet* has been rounded more than the roof of the hardtop coupe, the car is only 57 inches high. FORD ROUGE NEWS, DECEMBER 6, 1956

Family-owned since 1919, Ford went public in 1956. This photo of the promising new '57 Ford line was taken shortly after the historic event for the company's first ever Stockholder's Report. The stiffly-posed models were deemed necessary to add interest. Clockwise from the top right is a Lincoln Premier *4-Door Hardtop*, Mercury Montclair *Phaeton Coupe*, Ford Fairlane *Sunliner* convertible, Thunderbird, and Continental *Mark II 2-Door Hardtop*.

1958 THE THUNDERBIRD LOOK

SOMETHING went wrong in 1958. Not just for Ford but for GM, Chrysler and all the other automakers. Cars suddenly stopped selling just as they were being introduced, sending first Detroit, then Washington into a panic. Politicians were soon pleading for Americans to go out and buy a new car to help the nation's economy which was heavily dependent on the auto industry for creating nearly one in seven jobs and consuming a good part of the country's annual output, from steel and nickel to tires and radios.

Even Harley J. Earl, General Motors' god-like styling vice-president who always seemed to have the right answers, couldn't put his finger on it. Could it be that buyers didn't like what automakers were offering? There were complaints about Detroit's dream boats being too big, too bold, too gaudy, too loaded with chrome. Yet every automan knew from experience that flash, dash and dazzle — what automen called style — were attractions that sold cars.

But Harvard economist Sumner H. Slichter said Detroit had gone too far and partly blamed "the weird collection of headlights, fins, tails, wings, etc." for the recession. From Los Angeles Mayor Norris Poulson came criticism of the "big, flashy, chrome-encrusted, multi-powered car." *Time* magazine said 1958 car buyers were confused by the wild offerings of "jet intakes, bubble windshields, downswept snouts, and upswept fins and outswept taillights."

Chicago motivational researcher Louis Cheskin blamed it all on the space launch of the Russian Sputnik which happened about the same time the new '58 models were being launched. The reason cars weren't selling, he said, was . . . "The discovery that while we were fussing with useless decorations the Russians were making satellites and intercontinental rockets . . . (this) has had a profound psychological effect on almost every American."

From that observation, it couldn't have helped the sales outlook of Ford's new line of Edsel's that the very day they were introduced, the Soviets announced they had a missile that could drop a bomb anywhere in the U.S.A.

Joe Oros, head of Ford Division styling, was worried long before his new 1958 models ever had a chance to fall victim to the changing American psyche.

". . . My problem was what the '58 (Ford) should look like. Product planning wanted to change the '57 Ford car radically as a facelift from front to back, and they wanted a complete front end change, and a dramatic back end change, but, again, to be as dramatically successful as the '57. This was quite a full basket to request, and this proved to be an error . . . and we didn't quite make it, and, to me, that was a terrible agony.

It just was not as spontaneous a total theme as the '57. The '58 Ford front-end was borrowed from the (1958) Thunderbird front-end, but the proportions of the Ford car versus the Thunderbird proportions were radically different. The hood was so much higher than the Thunderbird hood from the ground. The front-end theme just didn't come off like a Thunderbird . . . The back-end, again, we tried to pick up the theme from the Thunderbird with the dual lights in the wide pod of the Thunderbird at the time."

Oros was also the main designer of the one Ford success story of 1958 — the new four-place Thunderbird which gave the '58 Ford its theme. On this note we can only imagine how the two cars might have looked if George Walker had prevailed. His candidate for the '58 T-Bird was a design that he personally helped Elwood Engle develop. According to stylist Gene Bordinat . . . "It was . . . a svelte, nifty looking automobile. It was rejected and the one Oros did was accepted . . . What were they going to do with it? . . . (Walker and Engle's car) . . . You know, with its very clean sides and bumper-in-grille kind of thing — *very nice automobile* — . . . It became the 1961 Lincoln."

There would be a lot of "what ifs" in the auto industry in 1958. When it was all over, because of changing attitudes and lost car sales, America had suffered its worst recession in the postwar era. Barely enough new cars were sold that year to replace the number junked — and Ford actually lost money, compared to record earnings in 1957. ◆

Left: **Out on a summer drive in their new '58 Fairlane "500"** *Sunliner* **Convertible. A surprising auto recession made it a rough year for the Thunderbird-styled Fords.**

NEW 1958 FORDS

Louisville Assembly Plant facilities entered the 45th year of Ford passenger car production on October 14th.

The 1958 Ford car line, featuring major styling changes, went into production here after one of the shortest down-time periods in Plant history. The Plant's production pace will be geared to 15 Ford cars and 20 Edsels every hour during the two eight-hour shifts.

Demonstrating the many 1958 Ford engineering changes are 21 models on two separate wheel-bases; the *Fairlane* and *Fairlane 500*, with an overall length of 207-inches, and the *Custom*, *Custom 300* and the station wagon, measuring 202-inches.

The new distinctive styling is emphasized in a front view of the massive wrap-around one-piece bumper with anodized aluminum "jet intake" grille, dual headlights, and *Power Flow* hood. New sheet metal treatment incorporates redesigned front fenders, new roof with seven front-to-rear flutes or grooves, and trunk lid and rear quarter panel innovations. From the rear the 1958 Ford is distinctive with a "V" sculptured trunk flaring into twin oval taillights.

Ford introduces the newest and most modern V-8 overhead valve engines in its V-8 line – the 332 and 352 cubic-inch engines. *Cruise-O-Matic*, an automatic transmission combining instantaneous "solid" response with nearly imperceptible up-shifting, is new for 1958. *Cruise-O-Matic* provides a new power train that combines overdrive economy with automatic transmission convenience.

The new Ford goes on display at Ford dealerships November 7th. LOUISVILLE FORD NEWS, LOUISVILLE, KENTUCKY, OCTOBER, 1957

Showing off her snazzy *Silvertone Blue Metallic Sunliner Convertible*. New this year was *Cruise-O-Matic*, *Even-Keel* suspension, *Magic-Circle* steering and "Safety-Twin" headlights.

Top right: A *Sunliner* Convertible provides a stage for a trio entertaining at a showing of new '58 Fords at the Shamrock Hotel in Houston. *Right:* A *4-Door Hardtop* at the same show. Round tail light theme was abandoned this year.

"The 1958 Ford still has its tubelike rear effect, and flaring canted fins, but other than that it is hardly recognizable, with a honeycomb jet-intake grille, dual headlights and spreading horizontal taillights." TIME

Sighed a Detroit secretary rapturously examining a trailerload of new 1958s: "Chrome is my favorite color."
TIME, NOVEMBER 4, 1957

At Dearborn Assembly the camera catches the details of a 1958 Ford "Six" engine just before the car's front-end is attached. Engine options this year were the 145 hp Six, and a choice of the 292 or the new "big block" 332 or 352 cubic-inch V-8s.

By 1958, with the buyer's wide range of choices in engines, colors, interior trim and options there were more than two million possible Ford car assembly combinations. Here, a worker at Dearborn pulls from a colorful array of headlight rims for the feeder line.

Right: Checking 1958 Ford *"Power-Flow"* hoods for factory-specified color match.

On the special Retractable line at Dearborn a '58 *Skyliner* ordered in *Colonial White* gets its intricate wiring hooked up. Hawk-eyed inspectors checked the top mechanism again at the end of the line where seasoned hands were stationed to fix anything from a bad relay to a crooked tail light. All Ford cars were shipped with the hubcaps stowed in the trunk to prevent damage and theft in transit.

Left: A Fairlane "500" *Sunliner* Convertible gets final line inspection and touch-up at Dearborn.

Nothing like it to impress the neighbors! A dazzling '58 Ford *Skyliner* Retractable puts away its top, ready for open air adventure. Introduced November 7, 1957, at a suggested list price of $2,907, it was top-of-the-line for Ford Division. Despite the auto recession a respectable 14,713 of them were sold.

"When the sun beams brightly... or the stars twinkle romantically... or your spirits quicken to the first soft breath of spring... that's when this car really comes into its own..." POPULAR SCIENCE

Some of the over abundant 18 models in the newly created 1958 Edsel line are pictured at the Ford Styling Rotunda. With names like *Ranger*, *Pacer*, *Corsair*, and *Citation*, a soon-to-be famous, sometimes made-fun-of grille, rocket lines, push-button drive and gull-wing taillights, they had enough to get people interested. What was missing were the customers eager to buy one!

Left: A '58 Edsel Citation *4-Door Hardtop*, wrapped for Christmas, wows children visiting the Ford Rotunda in Dearborn.

"What we wanted was for millions of people to be able to say at once: 'That's an Edsel.'"
GEORGE WALKER, FORD STYLIST

EDSEL MAKES DEBUT

The Ford brothers, Henry II, at the wheel, Benson and William, in a new Edsel Citation *Convertible*. Named for their father, the new Edsel line was created to give the company another nameplate to compete in the middle-priced field with such makes as Pontiac and Dodge. But as the brothers and their team of "experts" soon learned, a car such as the Edsel wasn't wanted — at least in the numbers needed to make a profit. The Edsel got off to a bad start in a bad year for selling cars and never recovered. Two years later it was quietly discontinued at huge losses to Ford. Saddest of all — the good name "Edsel" has become a synonym for failure.

A brilliant new vertical styling theme and several outstanding engineering innovations that include "Teletouch" push button transmission controls located in the steering wheel hub are features of America's newest automobile — the Edsel.

Available in 18 models and four series — *Ranger, Pacer, Corsair,* and *Citation* — the Edsel medium price car line offers two convertibles, sedans, two-door and four-door hardtops, and five station wagons. One convertible is available in the Pacer series and the other in the Citation, at the top of the line.

The Edsel vertical grille combined with an inner chrome impact ring and crisp horizontal sections on either side give an easily identifiable look of quiet elegance from blocks away. The concave sculptured sides have ever-widening teardrop effect and carry completely through to the taillights, giving a look of fluid motion and power.

The horizontal taillights blend smoothly into the flight deck luggage compartment lid to provide a solid bar of illumination on each side. Each bar is in two segments, divided at the luggage compartment lid. Outer segments contain turn indicators and brake warning lights in addition to normal red night lights.

Adding to the long, low look of the new Edsels is a slightly raised center section of the hood which recalls the elegance of motoring three decades ago.

The new E-400 Edsel engine is available in Ranger and Pacer series and the five station wagons. It develops 400 foot pounds of torque and 303 horsepower and with it comes a choice of standard, overdrive or automatic transmissions.

The E-475 engine, for the Corsair and Citation series, is rated at 475 foot pounds of torque and 345 horsepower. Only automatic transmission is available on these series. EDSEL DIVISION, PUBLIC RELATIONS, AUGUST 27, 1957

Ford Division general manager Robert S. McNamara, with top assistants, strikes a confident pose with his 1958 Ford line, including the new Fairlane, Thunderbird, Pickup and Truck. It would be a tough year for the man who would one day be Secretary of Defense under President Kennedy. The one triumph was the success of his controversial new 4-place Thunderbird.

NEW 4-SEAT THUNDERBIRD

Ford executives proudly introduce the all-new '58 Thunderbird at the Chicago Auto Show January 6, 1958. The news that Ford had dropped its sporty two-seaters stunned T-Bird fans. But it was strictly a business decision. Ford Division boss Robert McNamara had concluded that there was a far bigger market for a prestige family sports car with the same name. He was right. Sales doubled the first year and tripled the next.

"... I was surprised when they went to the family type 4-passenger Thunderbird ... Sales just took off like crazy ..." LEE KOLLINS, FORD ROTUNDA, PUBLIC RELATIONS

Ford ushered in the New Year last midnight by unveiling an automobile which may well change the American concept of luxury motoring.

It was the four-passenger Thunderbird which Ford chose to reveal before a group of prominent Americans during the New Year's Eve party at the exclusive Thunderbird Golf Club in Palm Springs, California.

The completely new automobile which will be introduced in Ford dealerships later this month is a stylist's dream car that combines the dash of the personalized Thunderbird with the interior spaciousness of a luxury car.

The Thunderbird retains the performance characteristics that made the two-passenger model famous the world over. It is powered with a new 300 hp engine of the most modern design, including a four-barrel carburetor, fully-machined combustion chambers and free-flow induction.

Inside is a console table running horizontally between the front seats. On it are the controls for the heater, air conditioner, power windows, a radio speaker, and ash trays for front and rear passengers.

Thunderbird styling features twin headlights deeply browed with the brow line extending into the hood. The functional air intake, midway back, raises the hood line to a widened sweep toward the windshield.

The flat roof line drops off to a novel rear window, yet the top retains the characteristic Thunderbird treatment in the rear quarter area. The rear trunk lid is deeply wrought with a wide valley separating the twin taillights which are set over a honeycomb pattern design similar to that of the anodized aluminum grille. *FORD NEWS BUREAU, JANUARY 1, 1958*

Left: Top Lincoln-Mercury men James Nance, left, and Ben Mills pose in a Continental *Mark III Convertible* at the Ford Test Track hill with their new 1958 models. They had their work cut out for them. It was already a bad year for the auto industry and their cars were part of the problem. Too big, too radically styled, too much chrome — all the things the American public rejected that year. When it was all over Mercury sales plummeted more than 50 percent and Lincoln fell from 41,123 cars sold in 1957 to 29,684 for '58.

Above: At $5,792 the Continental *Mark III Convertible* with leather upholstery as standard equipment was the only soft top in the new '58 Lincoln line which consisted of four models of Marks, and three each in the Premier and Capri series. Massive and wild styled, compared to the clean classic lines of the Mark II it replaced, with canted quad headlights, heavy front bumpers, and sculpted side body, the Mark III was controversial but actually sold better than the rest of the line. This year Lincoln went back to its roots and started all over again with a frameless unitized body like it had with the old Zephyr and first Continental.

"There is so much spendable income in the nation. We can't afford to relax for a moment. We must offer improvements and innovations just as quickly as we can perfect them."
HENRY FORD II, 1958.

1959 FORD AT FIFTY-MILLION

ONE thing learned from the 1958 auto recession was that buyers were becoming a lot more choosy. They still wanted youth, power, and performance but they were more likely to go for cars with toned-down styling than those with silly ornamentation and large amounts of flashy chrome.

Ford Division chief stylist Joe Oros, working under the direction of George Walker, knew that he had missed the mark with the design of his slow-selling '58 Ford models. He wasn't alone. So did designers of many other nameplates that year, like Buick and Oldsmobile. But with the '59s Oros felt he had found the right "rocket-styled" formula to make his cars more appealing:

"... On the '59 Ford we had our chance again because this was an all-new car ... all new sheetmetal with a very wide front end, hood-fender appearance, and, again the hood sunken down between the wide fenders similar to the '57, but the back end had a rocket theme derivative, and, again, aircraft influence.

... so, we had the rocket effect from the side beginning with the rear door and ending in the backup light. Underneath the backup light we had these giant taillights which must have been probably nine inches in diameter ... (and) ... were very dramatic. We treated the sheetmetal, the brightwork of the taillight, to give the effect of a jet turbine engine.

... That same year Chevrolet came out with its huge, gull-wing back and that was just too radical for the marketplace, and the people weren't quite ready for it. The Ford sales (department) had a field day describing this thing in the garage of the GM owners — wondering if it was a bug or a thing from Mars — and, of course, the '59 Ford that year outsold Chevrolet."

Some of the dealers who had seen advance pictures of

Left: Ford had come a long way in 56 years of auto production. Its milestone 50-Millionth is pictured at the Rotunda with one of the company's first cars, a 1903 Model A, and the futuristic "Levacar", as Henry Ford II makes the announcement that all three vehicles will go on a coast-to-coast publicity tour.

the new Ford before its introduction were very concerned about its huge taillights. "From a professional point of view," says Oros, "as far as design, there was nothing wrong with the aesthetics of the car ... They were large, but in total the taillights hung together very well with the sheetmetal. And, of course, after the introduction, all the dealers were pleased as punch with the car because it was such a great success ... That year was one of our banner years, and it was a wonderful comeback after the '58."

As business once again boomed, the 56-year-old Ford Motor Company — which was by now the third largest manufacturing company in the world — would celebrate one of the great milestones in auto making. On April 29, 1958, its 50-Millionth vehicle, a white Ford Galaxie *Town Sedan*, rolled off the line at Dearborn, Michigan. A month later, as a publicity stunt, it would set off on a cross-country tour to re-enact the famous 1909 New York to Seattle endurance race won by a Model T Ford.

Starting from New York City Hall June 1st, the history-making Ford would dip its wheels into the Atlantic and then follow the same basic route west to the Pacific as the Model T had in 1909. Leading an escort of company cars, vintage Fords, and vans carrying traveling exhibits, it would meander through the heartland of America, stopping at such places as Goodland, Kansas and Rawlins, Wyoming where whole towns turned out to see it.

Showcasing its 50-Millionth was one of the rewarding highlights of what would be a very transitional year for Ford to end an extraordinary decade.

Yet to come was the news in November that the Edsel was to be discontinued at heavy losses. And then, just as the company seemed to have found a winning combination that answered every type of public demand with more emphasis on the fast-selling station wagons and hardtops, a new kind of competitor was about to shake the industry — the small economy model foreign import car.

It was the wave of the future and Ford already had its own small car under development. On September 2, 1959, it introduced the compact new Falcon which would help change the attitudes of American car buyers forever. ◆

A loaded Fairlane "500" 4-Door Hardtop on exhibit at the 1959 Detroit Auto Show. Add-on accessories this year were the flashiest of the decade. Some of the choices were anodized side moldings, rear deck antenna, Sport Spare Wheel Carrier, "Flying Ellipse" hood ornament, Sun-ray wheel covers, fender skirts, stainless steel headlight "eyebrow" trim, and Deluxe Trim Package with "Flying Dart" rear quarter panel ornamentation, and "Tee Ball" fender ornaments.

THE 1959 FORDS

"... Daring and imagination wed Thunderbird elegance to the world's most beautifully proportioned cars... The car you'll most want to be seen in ..." FORD AD

Models at the Ford studios in Dearborn pose with a 1959 Ford *Sunliner* Convertible. The all-new '59 Fords were wider and longer than the '58s, with a big one-piece aluminum-starred grille, and canted tailfins. Glamour and fashion were the new advertising themes with lines like ... *The most becoming car to drive — sleekly beautiful ... luxurious as your favorite mink!...*

The 1959 Ford car line which already has received an award at the Brussels World's Fair for styling elegance, will go on display at Ford dealerships on Friday, October 17th, J.O. Wright, Ford Motor Company vice-president and Ford Division general manager, announced today.

"The 1959 Ford car is new in concept and new in style," Mr. Wright said. "The 1959 styling theme is good taste, a result of consumer demand for more elegance and dignity in automobile styling as opposed to gaudiness or extremism," he added. "Every piece of body sheet metal in the 1959 Ford has been changed to reflect the good taste of the Thunderbird — America's most successfully-styled car. The Thunderbird's crisp, taught lines have given the Ford a formal quality and the impression of a bigger, heavier car without the addition of bulk."

Ford's 1959 styling garnered the plaudits of the noted fashion authority, the *Comté Francais de l'Elégance*, which, for the first time in history, bestowed a gold medal for styling on an American automobile at the close of the Brussels international exposition.

The 1959 Ford line includes 17 models ranging from the *Custom 300's*, which are six inches longer than in 1958, through the six station wagon models to the *Fairlanes* and *Fairlane 500's*. All 1959 Fords will be on 118-inch wheelbase and will have an overall length of 208-inches.

The 1959 Ford with *Diamond Lustre* finish, a super enamel that doesn't require waxing, is available in a wide selection of solid and two-tone exterior colors, color-keyed to match the interior upholstery. FORD MOTOR COMPANY, DEARBORN, OCTOBER, 1958

FORD COLORS

Color has become one of the dominant features in automobile buying.

The parade of color starts on the branch feeder lines which deliver to the final line such components as wheels, seats, fenders, hoods, and tops – all painted to blend with predetermined color combinations.

The flow reaches a climax in the drive-away area, where hundreds of cars, with dozens of different color combinations, are stored ready for movement into dealer showrooms.

Selection of names for new colors has become an important element in automobile product planning. The names must have a pleasant association and sound for the car buyer.

Some of the Ford, Edsel, Mercury, Lincoln, and Continental color names for 1959 include *Satellite Blue, Petal Yellow, Jadeglint Green, Sunstone, Tawn, Moonrise Gray, Desert Tan, Tahitian Bronze, Geranium, Bolero Red, Autumn Smoke,* and *Twilight Turquoise.*

Colors used for all of Ford Motor Company's 1959 cars are "super-quality" enamel finishes. These newly developed synthetic-resin paints, used by Ford on a production basis since 1956, offer car owners many advantages over finishes of a few years ago.

The new enamels have far greater gloss, color retention, and durability. A mere washing brings out a renewed luster time after time. FORD NEWS DEPT., SEPTEMBER 29, 1958

A padded cradle riding on an overhead trolley brings a '59 Ford Galaxie *Sunliner* to its chassis. The man at the left readies the firewall which has all the necessary wiring, hoses, and attachments to make the connections. Men in pits along the line will tighten body bolts to the frame from underneath, and install other parts.

Right: "Decking" a front end unit on a 1959 Fairlane chassis at Dearborn Assembly, November 26, 1958. Buyers had three V-8 engine choices this year, ranging from the 292 rated at 200 hp, to the 332 Thunderbird V-8 with either 225 or 300 horsepower. The *Skyliner*, pictured with its lid raised, will leave this line once it gets its wheels and join other Retractables on an adjacent line to receive special work.

Last of the sun-loving, crowd-pleasing *Skyliner* Retractable series were the '59 models. An ad agency rolls out a strip of scrim to photograph a '59 *Skyliner* in its most famous pose — and sometimes its most infamous. Failure of the top's elaborate electrical system could happen at the most inopportune time. This problem led to steadily declining sales and the end to one of the most captivating American car series of all time.

Left: Women still had the most say when it came to buying an American convertible. Ford advertising shots show how good they'd look behind the wheel of a new '59 *Skyliner*, out in the breeze with the hardtop stashed away.

Right: A flashy Ford *Skyliner* gets its top mechanism tuned up at the 1959 Detroit Auto Show.

Left: Visitors to the Ford Rotunda in Dearborn admire a new 1959 Thunderbird *Hardtop*. Little changed from the hot-selling '58 models except for a horizontal bar theme in the grille and new side spear trim, the four-place "Bird" seemed destined for a long run.

Top left: At $3,979 the high-fashion new Thunderbird *Convertible* was just $283 more than the *Hardtop* model which outsold it about six-to-one. An option on either model was leather upholstery.

Above: The sleekly-styled new Mercurys at the 1959 Chicago Auto Show suggest new ideas in coming Fords, such as bigger windshields and more glass all around.

Above: Like the Skyliner, the 1959 Ford Ranchero was also the last of its kind and was destined to join the earlier '57-58 models as automotive classics. It was a good idea but the passenger car-styled pickup never sold well enough to justify its place in the Ford line. Production was stopped at the end of the 1959 model run of 14,169 units.

Right: As the decade came to an end, the Ford station wagon was still America's favorite hauler for town or out at the ranch. Of six models offered in 1959, including two kinds of Ranch Wagons and the pricier *Country Squire*, the top-seller this year was the *4-Door Country Sedan* model pictured here.

"The station wagon first started out as a farm carryall, then became a tricked-up luxury for the country club set. But today, by wedding the sedan to the wagon, Detroit's stylists have given it a new function; they have turned out a handsome auto that can be used either to haul tomatoes to market or top hats to the opera." TIME MAGAZINE

Top: At St. Louis, the half-way mark of its run to Seattle, Ford's 50-Millionth car and its escorts — including a sporty *Sunliner* convertible — are given an official welcome. Car No. 2 was a replica of the stripped down Model T that won the 1909 race.

Above: Townsfolk along the 1959 route admire the 50-Millionth, and its accompanying traveling exhibits — a fitting end-of-decade tribute to *"FORD'S GOLDEN FIFTIES."*

Right: A new Ford *Sunliner* with local beauty queens, a 1908 Model K Ford, and the 1909 Model T race replica, lead the 50-Millionth Ford and its caravan as it enters the outskirts of Seattle. The milestone car and its race reenactment were the Ford publicity highlights of 1959 as the curtain came down to end an extraordinary decade. ◆

PHOTO CREDITS

Except as noted below all photos in this book are courtesy of Ford Motor Company, Dearborn, Michigan.

Pages 79, 80, 102, 103, 113, 119, 125, 138, 139, 143, 161, Dan Brooks; 20, Tacoma Public Library; 21, 22, Huntington Library; 24, 152-153, 154-155, Henry Ford Museum; 31, 33 (top), 99 (bottom), University of Southern California Library; 30, Georgia State University; 33 (bottom), San Francisco Public Library; 86, 137, Florida State Archives; 63, Utah State Historical Society; 126-127, Mile Hi Photo Co., Denver; 183, Bob Bailey.